SHREWSBURY COLLEGE OF ARTS & TECHNOLC
MAIN CAMPUS, LONDON RD.

Telephone 01743 342354

D1589006

To Renew Book Quote No:
and last date stamped.

012914

Books are to be returned on or before
the last date below.

CD Rom avail

Printed Circuit Board Designer's Reference

Prentice Hall Modern Semiconductor Design Series

James R. Armstrong and F. Gail Gray
VHDL Design Representation and Synthesis

Mark Gordon Arnold
Verilog Digital Computer Design: Algorithms into Hardware

Jayaram Bhasker
A VHDL Primer, Third Edition

Mark D. Birnbaum
Essential Electronic Design Automation (EDA)

Eric Bogatin
Signal Integrity: Simplified

Douglas Brooks
Signal Integrity Issues and Printed Circuit Board Design

Kanad Chakraborty and Pinaki Mazumder
Fault-Tolerance and Reliability Techniques for High-Density Random-Access Memories

Ken Coffman
Real World FPGA Design with Verilog

Alfred Crouch
Design-for-Test for Digital IC's and Embedded Core Systems

Daniel P. Foty
MOSFET Modeling with SPICE: Principles and Practice

Nigel Horspool and Peter Gorman
The ASIC Handbook

Howard Johnson and Martin Graham
High-Speed Digital Design: A Handbook of Black Magic

Howard Johnson and Martin Graham
High-Speed Signal Propagation: Advanced Black Magic

Pinaki Mazumder and Elizabeth Rudnick
Genetic Algorithms for VLSI Design, Layout, and Test Automation

Farzad Nekoogar and Faranak Nekoogar
From ASICs to SOCs: A Practical Approach

Farzad Nekoogar
Timing Verification of Application-Specific Integrated Circuits (ASICs)

Samir Palnitkar
Design Verification with **e**

David Pellerin and Douglas Taylor
VHDL Made Easy!

Christopher T. Robertson
Printed Circuit Board Designer's Reference: Basics

Samir S. Rofail and Kiat-Seng Yeo
Low-Voltage Low-Power Digital BiCMOS Circuits: Circuit Design,Comparative Study, and Sensitivity Analysis

Frank Scarpino
VHDL and AHDL Digital System Implementation

Wayne Wolf
Modern VLSI Design: System-on-Chip Design, Third Edition

Kiat-Seng Yeo, Samir S. Rofail, and Wang-Ling Goh
CMOS/BiCMOS ULSI: Low Voltage, Low Power

Brian Young
Digital Signal Integrity: Modeling and Simulation with Interconnects and Packages

Bob Zeidman
Verilog Designer's Library

Printed Circuit Board Designer's Reference

Basics

Christopher T. Robertson

PRENTICE
HALL
PTR

PRENTICE HALL
Professional Technical Reference
Upper Saddle River, New Jersey 07458
www.phptr.com

Library of Congress Cataloging-in-Publication Data

Robertson, Christopher T.
  Printed circuit board designer's reference / Christopher T. Robertson.
      p. cm.
   Includes index.
   ISBN 0-13-067481-8
   1. Printed circuits Design and construction. I. Title.
   TK7868.P7 R63 2003
   621.3815'31--dc22

                                          2003019498

Editorial/production supervision: Jessica Balch (Pine Tree Composition, Inc.)
Cover design director: Jerry Votta
Cover design: Nina Scuderi
Art director: Gail Cocker-Bogusz
Manufacturing buyer: Maura Zaldivar

Publisher: Bernard Goodwin
Editorial assistant: Michelle Vincenti
Marketing manager: Dan DePasquale
Full-service production manager: Anne R. Garcia
Development editor: Jim Markham

© 2004 Pearson Education, Inc.
Publishing as Prentice Hall Professional Technical Reference
Upper Saddle River, NJ 07458

Prentice Hall PTR offers excellent discounts on this book when ordered in quantity
for bulk purchases or special sales. For more information, please contact:

U.S. Corporate and Government Sales
1-800-382-3419
corpsales@pearsontechgroup.com

For sales outside of the U.S., please contact:

International Sales
1-317-581-3793
international@pearsontechgroup.com

Printed in the United States of America
First Printing

ISBN 0-13-067481-8

Pearson Education Ltd., *London*
Pearson Education Australia Pty, Limited, *Sydney*
Pearson Education Singapore, Pte. Ltd.
Pearson Education North Asia Ltd., *Hong Kong*
Pearson Education Canada, Ltd., *Toronto*
Pearson Educación de Mexico, S.A. de C.V.
Pearson Education–Japan, *Tokyo*
Pearson Education Malaysia, Pte. Ltd.

This book is dedicated in the loving memory of my Father, Grandmother, and Grandfather, who all passed away during the writing of this book.

About Prentice Hall Professional Technical Reference

With origins reaching back to the industry's first computer science publishing program in the 1960s, and formally launched as its own imprint in 1986, Prentice Hall Professional Technical Reference (PH PTR) has developed into the leading provider of technical books in the world today. Our editors now publish over 200 books annually, authored by leaders in the fields of computing, engineering, and business.

Our roots are firmly planted in the soil that gave rise to the technical revolution. Our bookshelf contains many of the industry's computing and engineering classics: Kernighan and Ritchie's *C Programming Language*, Nemeth's *UNIX System Adminstration Handbook*, Horstmann's *Core Java*, and Johnson's *High-Speed Digital Design*.

PH PTR acknowledges its auspicious beginnings while it looks to the future for inspiration. We continue to evolve and break new ground in publishing by providing today's professionals with tomorrow's solutions.

PRENTICE
HALL
PTR

Contents

Preface

Printed Circuit Board Designer's Reference was written to provide a guideline of the entire PCB design/creation process, with reference material, software, forms, and other tools.

There are few books for the basic designer, and fabricators have limited published standards. A difficult decision about writing this book was publishing new, unreleased information that has few studies associated with it or little documentation supporting the values. History has shown that common practice and experience can suffice for undocumented information. Many designers and leaders in the industry feel that information shouldn't be provided without documentation. Most standards have not come from reading literature but from experience, common knowledge, and discussions.

PCB design is based on an ever-changing technology that requires constant updating and documentation. The values noted in this book are not supported by the Institute for Interconnecting and Packaging Electronic Circuits (IPC). These values are an average of fabricator's requirements and design requirements and are a cross section of personal values and those of hundreds of other designers. These values may change based on requirements, applications, changes in technology, and personal/company standards.

The values in the Manufacturing Technologies table (Table 2.1, pp. 18–19) are based on average values of manufacturers across the United States. They aid in choosing the manufacturer based on the design requirements and designing according to the manufacturer's capabilities. They are only a set of rules to guide the designer in choosing a manufacturer based on the design requirements and manufacturing capabilities.

SCOPE

There are countless styles, materials, types, and details to PCB design, but only the basics are covered in this book. An advanced version of this book would cover additional subjects such as bend/flex, controlled impedance, and exotic and special materials. This book deals with 50 to 75% of the designs done in this country.

Industry attention focuses on new and emerging technologies and techniques and ignores the basics of design.

Only the generic process and normal designs are covered in this book. Venturing into advanced board design requires a strong grasp of the basics of PCB design, which is the intent of this book.

One of the attempts of this book is to point out the weakness of the industry and the lack of complete software for the basic design. Daily tasks for most design issues are still not supported by many software packages; therefore, a supplemental software package accompanies this book that deals with the everyday tasks. A goal of this book is the education of software developers and the integration of these tables and calculators into PCB design software.

WHAT A DESIGNER SHOULD KNOW

A beginning designer needs only the knowledge of components and PCBs. This will allow the designer to design basic boards. To progress further, working knowledge of a computer-aided design (CAD) system and the essentials of electronics and electrical theory are necessary. These will allow you to have an understanding of "why a trace width must be this large" or "why the clearance must be so large."

To master PCB design, an understanding of RF and EMF is important.

There are many designers who do high-speed, high-frequency designs and who have no more than a high school education, and there are designers who have college degrees in engineering and who design simple boards. Many of the boards designed in the United States are low-frequency, simple boards and do not require advanced degrees to design.

One industry dilemma is the increase in design needs and a decline in the number of young/new designers. This decline is caused by increasingly complex software and more complex boards in the face of salaries that have not kept pace.

A second dilemma is the lack of standardization in the manufacturing industry. This causes confusion and disagreement among designers and difficulty in documentation.

A final dilemma is a lack of comprehensive PCB programs that account for real-life issues and the lack of understanding by the software designers of PCB requirements, such as calculations necessary on a daily basis:

> Instant trace width calculations, by layer, from current requirements
> Instant spacing calculations, by layer, from voltage requirements
> Impedance calculations by layer
> Layer calculations and documentations using real material values
> Database of available materials (derived from manufacturers)

HOW THIS BOOK IS ORGANIZED

An attempt was made to provide clear information concerning values, where they came from, and how to adjust them. Therefore, this book features the following:

> Easy-to-use tools for everyday calculations
> Easy-to-understand tables
> Quick reference charts
> A full checklist, beginning with the development and ending with final inspection
> Definitions, explanations, and graphics to clearly explain the numbers, values, and
>> results

As technology grows, standards and values will change; thus it is permissible to mark this book with newer relative values. The software is designed to grow with the technology and to provide the designer with a life-long design tool.

CONVENTIONS USED

This book contains many standards and conventions, which are detailed in this section. Most of the terminology is defined on the first use, but if not, is defined in the back of the book, depending on the point of view of manufacturer or designer. Some terms are redundant, and some definitions for the same word are different depending on the use.

Figures and Tables

Throughout this book there are tables and figures. Tables are defined as information provided in a columnar format, giving the reader an individual row for each item and an individual column for values. Tables differ from figures, which are pictures or graphic representations. The figures are sequentially numbered by chapter number and the number of sequence, similar to tables but separately.

Designer's Checklist

The Designer's Checklist is a comprehensive detail of the design process that may be used with every design and customized per the design's specific application (an electronic version is provided in the accompanying software). This checklist provides a proven structure while creating consistency from design to design.

Excerpts from the Designer's Checklist are integrated into the following chapters:

Chapter 4, "Schematics and the Netlist," incorporates the schematic input portion of the checklist.

Chapter 5, "Designing a PCB," incorporates the design portion of the checklist.

Chapter 7, "Board Completion and Inspection," incorporates the incoming board inspection portion of the checklist.

Chapter 8, "Drawing an Assembly," incorporates the assembly drawing portion of the checklist.

The checklist was integrated into the book to provide a guide for those checklist items in order of completion.

Notes and Tips

Throughout this text, there are notes and tips. These notes and tips are very important, and special attention should be paid to these items.

Metric versus Inch

Throughout this book inches will be noted and subsequently the metric equivalent will be noted in brackets, [ ]. One word of caution: Many standard values are based on the measurement system used; therefore, the metric equivalents may not be precise for metric designed boards.

The reality is that many components, systems, and manufacturers are still based on the inch measurements, and this can cause confusion. Fortunately, component makers and standards appear to be making the change to metric use. It is recommended for new designers to start with metric use and adjust measurements rounding to a common, even number.

SOFTWARE PROVIDED

At the rear of this book is a companion CD that contains many useful documents and program for the beginning user, including the aforementioned Designer's Checklist and the Designer's Reference Resource.

In addition, the software contains:

Borders in ANSI and metric formats
Drafting details
Drill charts
Sample fabrication notes
Example lay-up (stack-up) graphics
Title blocks in ANSI and metric formats
Other supporting viewers and programs

WEB SITE

Additional information and examples are provided in the support Web site for this book: *www.PCBDR.com*.

ACKNOWLEDGMENTS

Thanks go to the following:

Carol Robertson for preliminary reading and editing
Judy Eigelsbach for secondary reading and editing
James Markham for final review editing and finalizing the book and most accompanying materials
Bernard Goodwin for believing in me and publishing this book
God, for whom all things are possible

Designer's Checklist

Circuit Designer Requirements

Initial Planning

❒ Gather required information for part list, required component locations, and mechanical locations and requirements.

❒ Determine if all components are available in existing libraries. If not, use a component creation checklist.

❒ Select design template.

❒ Save file by part number.

❒ Enter design information.

❒ Open/load necessary libraries.

❒ Place components and wire together.

❒ Note all current, voltage, high frequency, noise and circuits.

❒ Add a note like the following for the most commonly used trace and space in the board designs: *"Unless otherwise specified, all circuits are less than .25A and 30V"* (this works for 6/6; .006" trace & .006" space).

❒ Place power-pin table.

❒ Place "last used" and "unused pin/gate" table.

❒ Highlight power nets and check each sheet for connectivity.

❒ Check for design rules, such as single node net; no node net; unconnected pins; unconnected wires; or other.

❒ Generate BOM and compare against part list.

❒ Add necessary notes.

❒ Add sheet numbers.

❒ Print and check schematic visually.

❒ Align/modify location, format and styles.

❒ Place all nets into classes.

❒ Generate netlist.

❒ Archive libraries.

❒ Check for other concerns:

    ❒ *Are all IC inputs terminated as required?*

    ❒ *Do IC/components have necessary filter caps?*

    ❒ *Are main circuit and branch circuits clearly identified?*

Printed Circuit Board Design Checklist

(This topic covered in Chapter 5)

Define Constraints

❐ Define board dimensions.

❐ Define top and bottom board clearance.

❐ Note dimensions of cutouts slots and unusable areas.

❐ Define the board thickness.

❐ Define edge clearance areas.

❐ Define all slots and cutouts.

❐ Define assembly requirement such as keying information.

❐ Mark predefined component locations including hardware, connectors, lights/LEDs, and switches.

❐ Place polygon on mask layer for area that requires no mask.

❐ Place keep-out (all layers or per layer) on area that require clearance from/for traces, vias, pads, and hardware clearance (by hardware-to-hole tolerance, hardware movement area, or hole tolerance).

❐ Define requirements: IPC, Mil-spec, etc.

❐ Determine assembly type for production, such as manual, automatic, or manual prototype-to-automatic.

❐ Determine servicing type:

  ❐ *No service/troubleshooting (throw away board)*

  ❐ *Low service (inexpensive components on the board, or easily swappable application, low serviceable location)*

  ❐ *Highly serviceable (expensive components on the board or difficult to swap application, highly serviceable location)*

  ❐ *Determine technology limitations and target technology.*

Begin Design

❐ Open new file or load appropriate template.

❐ Check for standards in pads, vias, or text styles.

❐ Draw board border using .040" line on center.

❐ Draw all slots/cutouts in board using .040" line on center.

❐ Enter design information.

❐ Load libraries/archived library.

❏ Load netlist.

❏ Generate BOM and compare against parts list. (This is to include mechanical components not in the schematic.)

❏ Place parts with a predefined location where necessary.

❏ Define classes/nets with trace width, clearance (space), and hole clearance

❏ Define other design attributes or design rules.

❏ If applicable, define class trace and space by layer.

❏ Configure design/job for:

    ❏ *Overall design rules*

    ❏ *Mask swell (global)*

    ❏ *Paste swell (global)*

    ❏ *Plane swell (global)*

    ❏ *Pad swell (global)*

    ❏ *Thermal divide (pad dia./4)*

    ❏ *Thermal clearance*

❏ Define assembly direction (especially auto assembly).

❏ Define/determine component direction.

❏ Define areas by type.

❏ Define layers including:

    ❏ *Number*

    ❏ *Symmetry (signal plane signal, etc.)*

    ❏ *Layer direction*

    ❏ *Layer type (strictly power, digital, etc.)*

    ❏ *Split planes*

❏ Copy board, cutout, slot outlines to all plane layer providing copper to edge clearances.

❏ Add text to plane layer as to net name (GND, +5V, etc.).

❏ Calculate board thickness and determine material availability. Attempt to use predefined or previously used combinations. Or, after design completion save successful stack-up combinations.

❏ Add tooling holes, if appropriate.

❏ Add datums to all layers or overlay layer. This helps not only to verify alignment after completion but for manufacturing alignment.

❏ Board part number in copper on bottom side

❏ Board revision (in copper or manually marking)

❏ Layer number (each layer, numbered by layer number, each offset)

❏ Assembly number (on silkscreen, top-side)

❏ Assembly revision (leave blank area for manual marking)

Initial Checks

❏ Check power pins are connected correctly on 1 of each type of part.

❏ Check plated-mounting holes are grounded when required.

❏ Complete placement location and prepare for routing.

Manual Routing

❏ Route the following types of nets first:

    ❏ *Most difficult*

    ❏ *Most complex*

    ❏ *Tight fitting nets first*

    ❏ *Very high current (primarily external)*

    ❏ *Very high voltage (primarily internal)*

    ❏ *Sensitive*

    ❏ *Noisy*

❏ Separate analog and digital.

❏ Route busses.

Auto Routing

❏ Manually route those items shown in "manual routing" first, if necessary.

❏ Define attributes that are common only to the auto router.

❏ Define/select "Routine," "Do" file, "Route" file, or "Strategy" file.

❏ After route completion:

    ❏ *Manually clean up paths.*

    ❏ *Miter right angle corners.*

    ❏ *Run DRC/design rules to ensure clearances are met.*

    ❏ *Check annular ring.*

    ❏ *Change gates or parts.*

Additional Markings

❏ ESD symbol

❏ High voltage warning

❏ Run DRC/design rules to ensure clearances are met.

❏ Relocate reference designators to their correct position/location/orientation.

❏ Relabel/renumber reference designator.

Creating a Manufacturing/Fabrication Drawing

❒ Copy border(s) to drawing layer or include border layer.

❒ Dimension the board in X and Y dimensions.

❒ Hole to edge dimensions (This is used for registration verification for Gerber/drill loading)

❒ Dimension and tolerance of any cutouts under +/- .005" tolerance.

❒ Board lay-up including the following:

    ❒ *Layer number*

    ❒ *Layer type*

    ❒ *Layer thickness*

    ❒ *Layer tolerance*

    ❒ *Copper layer type*

    ❒ *Min trace width spacing per layer (special cases only)*

    ❒ *Overall board thickness*

    ❒ *Overall board tolerance (Conventional +/-10%)*

❒ Drill legend, including:

    ❒ *Finished hole size*

    ❒ *Hole type (Plated or non-plated)*

    ❒ *Hole tolerance (Holes under .080" +/-.003". Holes over .080" +/-.005". Changes per technology)*

    ❒ *Symbol (Correlates with fabrication drawing or Gerber export)*

❒ Load or define fabrication notes including:

    ❒ *Guidelines or specifications to follow unless otherwise noted (PC class quality) and type SS/DS or ML)*

    ❒ *Material used (Core and pre-preg)*

❒ Is copper thickness specified per table?

❒ Min trace width and tolerance (+/-.003" general .001" tight)

❒ Min clearance and tolerance (+/-.003" general .001" tight)

❒ Plating per same table (more plating, more plating in hole, increased MFG AR)

❒ Hole plating minimum of .0002 (usually external plating)

❒ Finish type: HASL or tin lead (check for availability)

❒ Hole to pad registration (no breakout allowed)

❒ Layers to layer 1 registration (+/-.002")

❒ Overall scale tolerance (+/-.002 per inch. +/-.005" overall).

❒ Board size tolerance (+/-.005")

❒ Slot tolerance (+/-.003" to −.005")

❒ Beveling (if required)

❒ Electrical test and receipt of official test results. If required by P.O. or prototype only. Continuity of less than 5 Ohms per inch. Test at 100 V.

❒ One or more of the following manufacturing markings (usually placed on the bottom side):

    ❒ *Cage code (normally used by military contractors)*

    ❒ *Company logo (for identification if additional parts need to be ordered later in time)*

    ❒ *Date code (for board history)*

    ❒ *Lot code (for troubleshooting)*

    ❒ *Electrical test verification marking*

❒ Twist & bow value (.010" def., .007" tight)

❒ Coupon or x-ray inspection for hole wall quality (one of the most important quality aspects of a board)

❒ Other

Documentation

❒ Sheet/numbers of sheets

❒ Load or add information block specifying the following (this information may stay with the board until it is removed from the panel):

    ❒ *Company name*

    ❒ *Company phone*

    ❒ *Layer name*

    ❒ *Layer number*

    ❒ *Part number*

    ❒ *Revision*

    ❒ *Sheet of sheets*

Application Company Specific Information

❒ Add sheet revision block on first page (fabrication drawing).

❒ Add sheet revision section (border information).

❒ Update design information such as the following:

    ❒ *Date (update every time this file is finished, changed, or modified)*

    ❒ *Designed by (designer name)*

➗ *Engineer (electrical engineer or the schematic's entry person)*

➗ *Checked by (QC, final, or engineer's name)*

➗ Add sheet revision block on first page. (fabrication drawing)

Check Plots (not required)

➗ Print each layer w/o boarder to scale.

➗ Inspect for the following:

    ➗ *Sheet layer numbers*

    ➗ *Datums*

    ➗ *ESD symbol*

    ➗ *HV note*

    ➗ *Tooling*

    ➗ *Pin 1 identification*

    ➗ *Mounting hole locations*

    ➗ *Board size and clearance*

    ➗ *Mechanical support*

    ➗ *Hardware clearance*

    ➗ *Stack-up thickness*

Approval

➗ Get PCB approval from engineer.

➗ Implement any redlines.

➗ Generate netlist from schematic again.

➗ Run DRC again and run compare netlist.

Output

➗ Set up Gerber output files or set up database export.

➗ Export the following (in 274-X):

    ➗ *All layers separately*

    ➗ *Required silk screen layers*

    ➗ *Top and bottom solder mask separately*

    ➗ *Fabrication drawing with symbols*

    ➗ *Drill file (in ASCII format, leading suppression)*

➗ Load Gerbers in a CAM/CAD viewer and inspect for consistency with original design.

File Archive

❐ May be done after receipt of board or when test is complete.

❐ Place files in restricted area and change file properties as Read-only.

❐ If changes need to be made, change revision and start a new file using revision descriptor in the part number.

Incoming Board Inspection

(This topic covered in Chapter 7)

❐ Check initial look of board. Note cleanliness and appearance.

❐ Check mask, including:

  ❐ *Specified color*

  ❐ *Specified thickness*

  ❐ *Quality*

  ❐ *Blemished*

  ❐ *Pitting*

❐ Is plating adequate?

❐ Are holes centered in pad (annular ring and alignment)?

❐ Are hole sizes correct?

❐ Are layers registered?

❐ Adhesion test

❐ Does appearance match artwork?

❐ Does silkscreen match artwork?

❐ Are overall board dimensions correct?

❐ Is board warped?

❐ Is any copper showing on board edges?

❐ Trace width in tolerance

❐ Check electrically the following items:

  ❐ *Several plated holes front to back (less than 2 Ohms)*

  ❐ *Longest trace from one end to the other (less than 5 Ohms per inch, or specified value)*

  ❐ *Resistance between planes of different nets for shorting (should be open)*

  ❐ *Test pads closest place for continuity (known opens should read open)*

Printed Circuit Board Assembly Checklist

Create an Assembly Drawing

(This topic covered in Chapter 8)

❐ Load title block on top assembly layer.

❐ Add page # to each sheet.

❐ Enter design info on assembly fields or designated areas.

❐ Load basic notes according to board type.

❐ Load silk Gerber, remove all exterior to the board and copy to top assembly.

❐ Draw side/bottom view of board (including parts and screws, etc.).

❐ Using parts list, add find number leader with quantity.

❐ Add notes as required and triangle find numbers as needed.

❐ Place additional tables for any wire(s) used. Always add note and triangle find number with basic note.

❐ Check for mounting hardware on all components and sub-assemblies.

❐ Check soldering, shrink (all soldered terminals), glue locations, required fixture, mounting brackets, and card guides.

Conformal Coating only

❐ Place phantom lines.

❐ Place dimensions for conformal coat clearance with note find number.

❐ Place serial number box block next to the serial number silkscreen.

❐ Add appropriate additional notes and place find numbers

❐ Place appropriate distribution statements and proprietary notes

❐ Place revision sheet information on all layers.

❐ Place revision status block on assembly drawing, top sheet.

❐ Get assembly drawing(s) approval.

❐ Compress Gerbers into *drawing number*.ext.

Introduction to a Printed Circuit Board

This chapter serves as an introduction to printed circuit boards (PCBs). It provides the basic understanding of a PCB and details the materials, objects, and forms that a PCB requires. Information gathering and design constraints are also detailed. This ensures a well-rounded understanding of all components required.

Basic understanding of the materials and terminology of a PCB is critical in clearly conveying requirements and intent. Later chapters will cover these objects in more detail.

WHAT IS A PCB?

A PCB is used primarily to create a connection between components, such as resistors, integrated circuits, and connectors. It may also be used in the following items:

- A keyboard. When a button is pressed, a plastic bubble, with a metallic piece on the underside, is depressed and bridges two traces or contacts, thus completing a circuit.

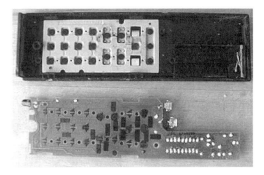

Figure 1-1   Multimeter enclosure face and internal.

Figure 1-2   Remote PCB.

Figure 1-3   PCB inside a computer.

Figure 1–4   Automotive computer.

- A multimeter. This has a PCB that must be aligned with a switch and has a cutout for the meter, as shown in Figure 1–1. It also has holes and jacks for the leads, uses metal tabs for the battery, must conform to the size/outline of the case, uses a component that mounts through the board, and can be adjusted from outside the case.
- A remote. A combination of mechanical components that interface with a PCB to work other electronics that contain PCBs (Figure 1–2).
- A computer. One of the most complicated boards that supports processors, ICs, as well as connectors for other boards (Figures 1–3 and 1–4).

WHAT A PCB IS MADE OF

A basic PCB starts with a copper-clad fiberglass material or thin copper sheets adhered to either side of the board, as shown in Figure 1–5.

With a multilayer board (a board with more than two copper layers), a piece of Pre-Preg (Figure 1–6) may be placed between these cores to create one solid board with several copper layers (Figure 1–7). Pre-Preg is made of material similar to the core with additional adhesive that will adhere it to the layer above and below.

To discuss the board materials and options to a manufacturer, the terminology should be understood. The following sections explain the materials and the items that make up the material.

Core/Core Material

Core material (see Figure 1–5) is a rigid sheet of fiberglass resin material that has two sheets of copper adhered to either side. Some material may have a copper

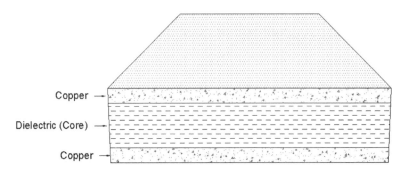

Copper

Dielectric (Core)

Copper

Figure 1–5   Core material.

sheet on only one side. The copper is measured in ounces (oz). The normal thickness, by ounce, is as follows:

- 1/2 oz   (.0007″[0.01778])
- 1 oz   (.0014″[0.03556])
- 2 oz   (.0028″[0.07112])
- 3 oz   (.0042″[0.10668])

PCB manufacturers will refer to the copper thickness in ounces, but during board lay-up, or when the materials are stacked together, the inch/mm thickness is used.

Pre-Preg

Pre-Preg material is made of similar material as the core material but is in a soft, pliable form and comes in standard-sized thin sheets. These sheets are stacked to create the following custom thickness (see Figure 1–6):

- .002″ [0.0508]
- .003″ [0.0762]
- .004″ [0.1016]

Dielectric (Pre-Preg)
Dielectric (Pre-Preg)
Dielectric (Pre-Preg)

Figure 1–6   Pre-Preg material.

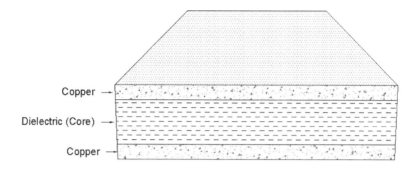

Copper →

Dielectric (Core) →

Copper →

Figure 1–7   Core material with Pre-Preg material creating a single board (one of two types).

When thickness calculations are made, multiples of .002″[.0508] or combinations are recommended.

Copper Foil

Copper foil (Figure 1–8) is a thin sheet of copper that is placed on or between Pre-Preg materials and bonds to the Pre-Preg with the adhesive that is part of the Pre-Preg. These materials come in the following thickness:

- 1/2 oz   (.0007″[0.01778])
- 1 oz   (.0014″[0.03556])
- 2 oz   (.0028″[0.07112])

Copper Plating

Copper plating is primarily used only on the finished board, on the external layers, and provides an additional thickness of copper to the board while plating the wall of the holes drilled in the board.

The hole wall plating is the main purpose for the copper plating but still must be added into the overall thickness of the board. The average thickness for the plating is .0014″[.0356], ranging from .0012 to .0014″[0.0304–0.0356].

Copper Foil →

Figure 1–8   Copper foil.

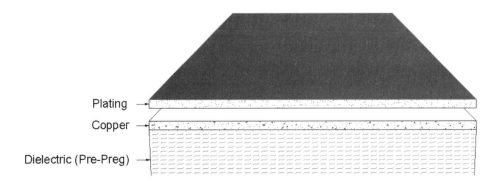

Plating
Copper
Dielectric (Pre-Preg)

Figure 1–9   External plating.

Usually the external plating is added after the board is drilled and the external copper on the board is etched, leaving a thicker trace, as shown in Figures 1–9 and 1–10.

Solder Flow

Solder flow is a process in which solder is applied to the external surfaces of the board on exposed copper areas (Figure 1–11). This helps prepare the board for soldering and protect the copper from oxidation. The copper areas on the entire board may be solder flowed, or a process called SMOBC (Solder Mask Over Bare Copper) will be used. SMOBC process is when the board is "masked" and only the exposed area (usually pads or areas that are to be soldered) will be coated with solder flow.

Solder Mask

Solder mask (Figure 1–12) is a material that is used to coat the board to

- Protect from surrounding environment.
- Insulate the board electrically.
- Protect against solder bridges.
- Protect components mounted to the board.
- Protect the board from heat generated from components mounted to the board.

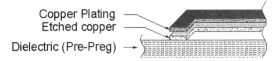

Copper Plating
Etched copper
Dielectric (Pre-Preg)

Figure 1–10   Plating after etching.

Solder Flow
Copper Plating
Etched copper

Dielectric (Pre-Preg)

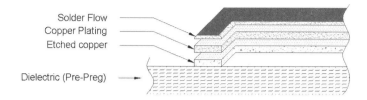

Figure 1–11   Solder flow.

Some standards do not view solder mask as an adequate insulator because of the inconsistency between board houses in types and inconsistent thickness after application.

The Trace

A trace on/in a PCB is relative to a wire. It provides the same function of transferring electricity from one point to another while the strength of the PCB provides a rigid material to place components on.

The copper layers, as explained later, will each go through an etch process that removes unneeded portions of the copper, leaving only those traces and pads required.

The defining characteristics of a trace are the trace width and height/thickness. These are the values that determine the amount of current that a trace can handle, similar to wire. The rule of thumb for trace width is .010″[.0254] per 1 Amp external (normally 1.5 oz or .0021″[.0533]) and .040″[.1016] per 1 Amp internally (for 1/2 oz of copper). *This is only a guideline.* The actual formula is much more complex because it does not follow a standard multiplier. The values mentioned will guide the work but do not consider temperature and copper thickness. (The accompanying software includes a calculator that takes in all values to consider.)

- Temperature (increase above room temperature)
- Current
- Trace width
- Internal/external layer

Solder Mask
Finished Trace

Dielectric (Pre-Preg)

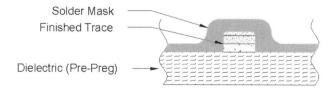

Figure 1–12   Solder masking.

Note

Formulas are necessary, but without a proper utility they are difficult to use and time consuming, which is why the accompanying software supports most formulas given in this book.

The Pad

A pad may consist of several different shapes and styles. Normally two types of pads are used, commonly referred to as a soldered surface-mount pad or a soldered thru-hole pad (Figure 1–13).

A surface mount pad is nothing more than a square or rectangular copper area that is used for mounting surface-mount components. The size and shape depend on the component that is mounted/soldered to the pad. Most component manufacturers have recommended pad sizes for their components.

A soldered pad may consist of a plated thru-hole pad (PLTH) or a nonplated thru-hole pad (NPTH). Both are nothing more than a round, square, or oblong pad

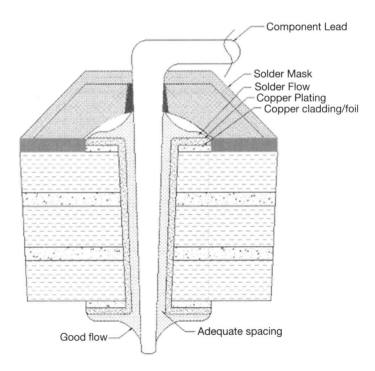

Figure 1–13   Lead soldered into a board.

with a hole through the pad. This allows a leaded component to be mounted to the board by placing the lead through the hole and the lead soldered to the pad area.

The Plated Hole

The plated hole, shown in Figure 1–14, consists of a pad in almost any shape with a hole through the pad. The walls of the hole are surfaced or plated with copper and, in some cases, solder or some other protective plating. The plating in the hole extends from the external area and flows into the hole, "plating" the hole wall.

A plated hole wall is superior for several reasons:

- Support of the external pad, allowing a smaller pad
- Dissipation of heat during soldering, allowing a smaller pad
- Connection from the top to the bottom layer regardless of component insertion
- Support solder flow from the top to the bottom, eliminating soldering from both sides
- Necessary for multilayer board for internal pad connection

The Non-Plated Thru-Hole

The non-plated thru-hole (NPTH) is nothing more than a pad with no plating in the hole. The pad is commonly used for single-sided boards or holes that have screw/mounting hardware. Figure 1–15 shows an unplated hole with no clearance on the inner copper layers. Unplated holes will commonly have an area around the hole clear of any copper (similar to the board edge) to prevent shorting

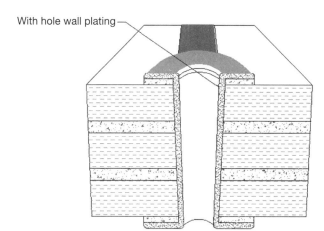

With hole wall plating

Figure 1–14   Plated hole cross section.

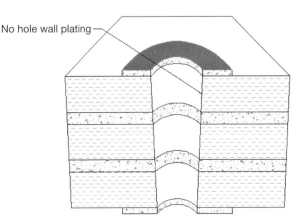
No hole wall plating

Figure 1–15 Non-plated hole cross section (no copper clearance).

between any objects that may be placed through the hole to any of the copper layers. If a board has no plated holes, it does not require the plating process. This reduces the overall cost (but in turn increases pad size), reduces available surface area, and doesn't allow internal layers.

Slots and Cutouts

Slots and cutouts are similar to plated/non-plated thru-holes but are defined separately because of their shapes and characteristics. PLTH/NPTH are usually round holes but not always round pads, but the hole is drilled. A slot or cutout is cut into the board by a router bit and is an oval, oblong, or a rounded square shape. The "corner" of a slot or cutout is not square unless a special corner punch is used. Slots and cutouts are cut with a round bit; therefore the corner must be the radius of the bit, or larger. The router bit size depends on board thickness and manufacturer's capabilities. A small bit may be used on a thick board but must cut slowly to prevent breakage, increasing time and cost.

The Board Edge

The board edge deserves its own definition and attributes, due to the values that define the board edge. The board edge is any part of the board that exposes the cross section of the board, including slots, cutouts, and the outer edge of the board.

A THUMBNAIL SKETCH OF THE DESIGN PROCESS

The entire design process is determined by the application and the designer-defined role. The design process can be defined as simply designing a board to provided specifications, or as a process of defining the environment, enclosure,

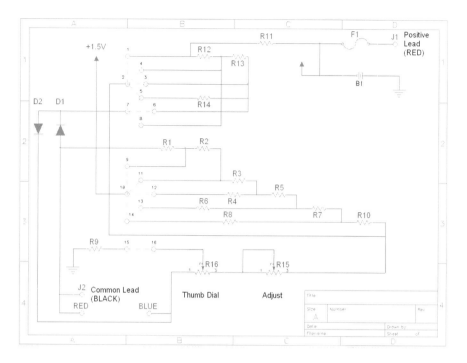

Figure 1–16   Schematic-capture entry.

Figure 1–17   Multi-meter in its enclosure.

connections, and mounting and implementing those requirements into the board design. This book takes into account many aspects of the design process, and this section defines some of the information that should be gathered.

First, the determination of function or form is made. Is there a size constraint for the board? If not, then the size of the board is determined by the components in the circuit and the area they consume.

This gives the engineer an idea of what he or she has to work with. The engineer will then create a circuit or circuits and enters them into an intelligent schematic capture program with results such as Figure 1–16. This then represents the completed board shown in Figure 1–17.

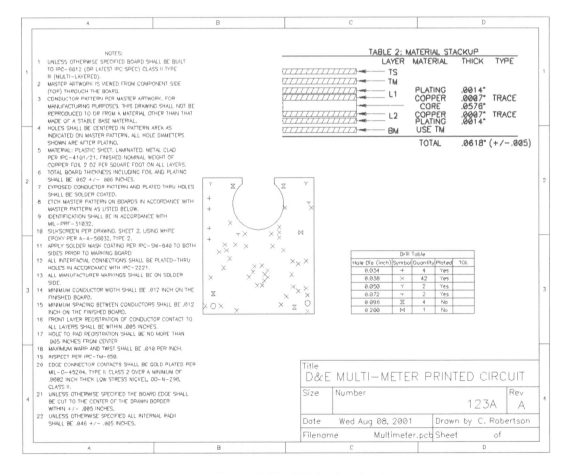

Figure 1–18   Fabrication drawing.

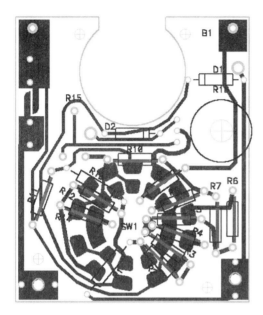

Figure 1–19   Gerber layers overlaid.

At this point the schematic data are captured and read into a CAD/CAM program, placing components whose schematic symbols were connected in the schematic capture program. The components are arranged, as necessary, and connected by traces following standards, manufacturing constraints, and design constraints.

The manufacturing specifications should then be noted and any constraints specified (Figure 1–18).

The data are transferred to the manufacturer, depending on their capabilities, quality, and, if required, their speed. Figure 1–19 is the combination of the two Gerber files sent to the manufacturer. A Gerber file is an electronic format representing an image of the board, usually one file for each layer of the board. The combined Gerber files create a data graphic file that represents the completed board.

The manufacturer then takes the designer's data, selects the required/specified materials, and processes the board, creating a bare board similar to the designer's drawings.

Sometimes components need to be mounted on the board in an automated fashion. The board(s) and components are sent to an assembler, where the components are then mounted to the PCB.

SUMMARY

This chapter discussed many of the applications of a printed circuit board and what a PCB is and does. The designer was shown the entire spectrum of objects, limits, and constraints regarding the general size and thickness of a PCB. After gaining a basic understanding of applications and limitations, some of the specific details and terminology of a PCB were outlined. Each of these items, seemingly simple, plays an important role in the board and requires individual attention and consideration.

With this information the designer can proceed to the manufacturing process, detailing how the board is actually manufactured and how to address each process.

2

Design for Manufacturing

This chapter will address the fabrication process of the PCB and the requirements of the manufacturer. Manufacturers are separated by their limitations or constraints into categories known as "technologies." These categories are determined primarily by cost. As the level of technology increases, so does the cost. These technology categories help designers control cost by limiting their designs.

This chapter explains the differences in the technologies, defines the limits, and details the step-by-step process, specifically of the conventional process and how the designer should write fabrication notes (instructions) for each process.

Each process is explained so the designer will understand the basics of how the process works and thus be able to make an educated change to notes when necessary.

Note

Throughout this chapter you will see many variations of the words *manufacture* and *fabricate*. For the purposes of this discussion, a manufacturer is a business that fabricates a board.

ABOUT FABRICATION NOTES

A designer's fabrication notes are a collection of notes that accompany PCB data files (Gerbers or some other data file) as a text file, or are provided as in a drawing of the PCB itself that conveys the designer's requirements and details the fabrication process. The fabrication notes are one of the most cryptic and confusing parts of the PCB process, and many designers are not sure how or what to specify in them. The notes are made even more difficult by the inconsistency of the manufacturer's requirements and the lack of guidelines. Before a designer can direct a manufacturer how to proceed, the designer must ask a few questions and understand the process.

Fabrication notes aren't made to restrict the manufacturer but to provide consistency and a starting point, which is important when attempting to adjust any values. The values specified in this chapter are based on conventional technology.

Note

There are two separate sets of specifications in this industry: *finished PCB specifications* and the *fabrication specifications*.

Finished specifications are those values specified by IPC, the Mil-specs, and/or UL™. (UL is the trademark of Underwriter's Laboratory, an organization that creates and test for standards in the electrical/electronics industry.) These are requirements set in place to ensure quality and consistency.

Fabrication specifications depend on the technology that they use and may vary from manufacturer to manufacturer. These specifications are a combination of fabrication limits and requirements necessary to achieve the finished specs. Many of the finished specs are based on these fabrication values (such as annular ring), and everything about the PCB is limited by these values (such as trace and space). The designer must first learn the limits of his or her design. These limits are controlled by the technology.

TECHNOLOGIES

Technology is the knowledge of how to create, produce, or perform some object or function. In PCB design the term *technologies* is no more than a categorization of values or capabilities of a manufacturer. These values are based on capabilities of the manufacturer's equipment and the overall process.

The three controlling points are *etch*, *drill*, and *registration*. Other capabilities influence the overall category, but these are the most important.

Previously, these technologies have not been clearly defined. Manufacturers have not bought into a category for fear of scaring off customers and displaying too much information for competitors to see. There are also no organizations or groups that record and organize such values. Therefore, during the creation of this book, a survey was taken of many PCB manufacturers, and the following categories were defined more clearly: conventional, advanced, leading edge, and state of the art (refer to the following section). As with all technologies, the values will change through time, and additional categories will evolve. These are the categories and their general definitions:

- **Conventional**—This is the lowest technology and is the most common. The general limitations of this technology are trace/space of .006″/.006″ (for .5 oz copper), a minimum finished drill of .012″[.3048], and 8 to 10 layers maximum.
- **Advanced**—Advanced technology is a higher level of technology, limited to 5/5, a minimum finished drill of .008″[.2032], and 15 to 20 layers.
- **Leading edge**—Leading-edge technology is essentially the highest level of manufacturing that is commonly used. This technology is limited to about 2/2, a minimum finished drill of .006″[.1524], and about 25 to 30 layers.
- **State of the art**—State-of-the-art technology is not well defined because it is an ever-changing technology whose values will change with time and must be adjusted regularly.

Note

Most general specifications in the industry are based on conventional technology as well as .5 oz starting copper.

DEFINING FABRICATION LIMITS

Table 2–1 (extracted from a survey conducted of many major U.S. PCB manufacturers) provides a list of fabrication specifications to start with when designing a PCB. The values shown are the minimum values or the limits of the manufacturer's capability. The designer should *not* use these values unless necessary. It is advisable to leave room for change and not always force the manufacturer to the limits.

On the accompanying CD, there is an editable table that a designer may use to record/update these values.

Table 2–1   Technology Table (Conventional)

Type	Conventional (.006/.006, .010)	Note
Etch back	0.0014	Per 1 oz copper (.0007 per copper side)
		Starting copper thickness
Trace min. .5 oz	0.0060	Per starting copper
Trace min. 1 oz	0.0070	
Trace min. 2 oz	0.0090	
Trace min. 3 oz	0.0100	
Space min. .5 oz	0.0060	Per starting copper
Space min. 1 oz	0.0070	Per starting copper
Space min. 2 oz	0.0090	
Space min. 3 oz	0.0100	
Trace tol. (±) per oz	0.0020	
Min. external copper (oz) starting	0.5	Starting copper/starting copper + plating
Max. external copper (oz) starting	3	
Min. internal copper (oz)	0.5	
Max internal copper (oz)	2	
Plating min	0.0014	Starting/plating (oz) 1/2oz = .0007″
Finished Drill Aspect	5 : 1	Board/drill (.062 board = .008 drill)
Min. drill (drilled hole)	0.0100	Finished drill = min drill – "drill over" (*unless plating thickness is increased*)
Drill tol. PLTH < .080	+/-.004	
Drill tol. PLTH	+/-.005	
Drill tol. NPTH	+/-.003	
Drill tol. NPTH	+/-.004	
Board edge clearance (plane)	0.020	Clearance for route
MFG AR total ML	0.009	
MFG AR total SS/DS	0.006	

Type	Conventional (.006/.006, .010)	Note
Image tolerance	0.003	
Drill tolerance	0.003	
Stack-up tolerance	0.003	
Min. Pre-Preg material thick	0.002	Thickness of a single sheet of Pre-Preg
Min. Pre-Preg Material thick	0.005	Thickness of a single sheet of Pre-Preg
Board max.	0.125	Thickness (FR4)
Board min. (DS) (core)	0.008	Excluding plating
Min. ML	0.020	Excluding plating (4layer)
Min. thick ML tolerance (± 10)	10	%
Min. material tolerance	0.02	Standard Fr4
Board max.	16x22	X" / Y"
# of layers	6	Maximum

Note: AR—annular ring; SS—silk screen; MSK—Mask. Contact your manufacturer for specific values/limits.

The following are some of the key terms and values, per technology, that limit a design. These terms are used throughout this book and within the industry.

- **Min. trace**—The minimum trace width is determined by technology and copper thickness. Listed is the most common thickness used. *Note*: These values are for the "starting" copper thickness.
- **Min. space**—Min. space is determined by the same values as min. space, except the min. space requirement increases as copper thickness increases.
- **Aspect ratio**—This is a proportional value. The first value is basically a divisor of the second. For example, 8:1 means a .064"[1.6256] board is divided by 8, giving a drill size of .008"[.2032]. A drill for a .125"[3.175] thick board can be no smaller than .015"[.381]
- **Min. drill**—Manufacturers have a limit on the size they may drill. This value is the smallest drill available to them, or they can drill with consistency.
- **Drill tolerance**—Drill tolerance is one of the defining factors of the manufacturer's technology. A hole is usually not perfect, for several reasons. Drill tolerance is the range that a specified drill many finish at.
- **Min. AR PLTH**—Regardless of specifications/requirements for the completed board, the manufacturer has a minimum requirement that must be

added to, or added into the designer's annular ring. This value is for the Plated Thru-Hole.

- **Min. AR NPTH**—This is essentially the same as description as the PLTH, except that it is larger, as explained later.
- **Mfg AR**—The additional amount of area beyond the designer/standards of a pad required by the manufacturer to compensate for errors during the process (image, drill, and layer registration).
- **Hole wall (plating)**—After drilling a board, the board is placed in a bath of water and copper flakes and electrically charged. The charging of the board attracts the copper and causes it to adhere to the copper around a hole and "grows" inward, creating a sleeve through the hole.
- **Copper plating**—Occurs during the hole plating, where copper attaches to the remaining exposed copper area. Copper plating is basically a feature of the hole plating.
- **Mask min. clearance**—This is an area around a pad or hole to be void of solder mask to account for mask registration errors.
- **Mask registration**—Mask registration is the location of the mask in reference to the image, data of the top layer, or holes in a board.
- **Mask avg. thickness**—If mask is being used as an insulator, the thickness must be maintained.
- **Mask min. thickness**—This is the minimum thickness manufactured with consistency.
- **Silk screen registration**—This is the registration, usually measured from the top layer to the silk screen.
- **Silk screen height**—The height is measurement of how tall the text is or the required height.
- **Silk screen thickness**—The thickness is the line width of the text or "stroke width."
- **Min. route**—This is the minimum bit size available, causing any slots to be at least this width or larger.
- **Min. radius**—Minimum radius is normally 1/2 of the minimum route or bit width. This measurement is for internal corners. All internal corners are limited to the minimum radius.
- **Bow and twist**—Bow and twist, sometimes known as warpage, is the amount of raise from a true flat surface to the board, per inch.
- **Board thickness tolerance**—Due to the cumulative value of materials creating the board and errors in the multilayer press process, the thickness may vary.

THE FABRICATION DRAWING

To provide the PCB manufacturer with a clear description of the requirements and the limitations of a design, the designer should supply the manufacturer with fabrication notes and a fabrication drawing (Figure 1.18, p. 12), along with all data files. These notes may be supplied in paper form, but a Gerber format (an electronic drawing format) is preferred so it may be maintained with the customer's files. The fabrication notes should contain the following essential items:

- Table/legend of drill sizes, tolerances, and quantity with a symbol legend.
- A drawing of the board with the respective symbols representing the holes in the board.
- Graphic representation of the board cross section detailing the layer number, type, and thickness.
- Manufacturing notes specific to only the manufacturing process and not the assembly process.
- Additional information such as revision and data and company information should also be placed on this drawing as well as any other internal tracking data.
- Detail of all cutouts, slots, and notches that are to be created during the manufacturing process.
- Outline or border of the board.
- Dimension, in x- and y-coordinates of the overall board.
- A hole to edge dimension for one hole near the edge, in the x- and y-coordinates for drill alignment verification. (If there are no holes in the board, this is not necessary since there are no holes to align.)

A board may be manufactured without any of this information, but final product may not be what was desired by the designer. If no information is provided, there is no recourse if a board does not meet expectations. Some designers forgo the actual outline and dimensions of the board and add a note for the manufacturer to follow the data of the Gerber and drill files. This adds more risk since there is no visual conformation of the data, and the Gerber files, which are a product of the design, may have discrepancies. Fabrication notes are also optional, but they provide a useful guideline for the manufacturer as well as documentation for the designer of how the board was manufactured.

The following are some common-sense guidelines to follow when creating fabrication notes:

- Notes should be clear and concise.
- There must be no repetitive notes or values that may contradict each other.
- Values specified are within the capabilities of the manufacturer.
- Notes are in order of the manufacturing process, to provide a checklist for the manufacturer.
- Notes should be thorough and limiting but not too intrusive to the process. Too many limitations and intrusive "process defining" notes will increase cost and increase the turnaround times.

Shall indicates a must, and *may* indicates an optional specification. Most notes that are generated may be used for future boards. Therefore, a designer should save these notes individually or in a collection, such as a large text file, based on the technology.

THE FABRICATION PROCESS AND FABRICATION NOTES

The following sections cover the fabrication notes or the fabrication process and fabrication notes used to specify parts of the process. There are many steps in the fabrication process, each requiring individual specifications. The following sections detail each process, mention specific values that need to be defined, and provide a sample note to use. For each section, the following information will be provided (if applicable):

- **Define** (The specific points that need to be defined)
- **Sample notes** (A sample note that the designer may use)

The following are the basic steps of PCB design. *(Some steps may need repeated, added, or removed per manufacturer.)*

1. Set-up
2. Imaging
3. Etching
4. Multilayer pressing (multilayer boards only)
5. Drilling
6. Plating
7. Masking
8. Board finishing
9. Silk screening

10. Route

11. Quality control

12. Electrical test

Set-Up

Set-up is the initial selection and determination of materials, processes, and requirements. It is the one process that does not deal with the physical board but rather determines what materials are used, how many materials are used, in what order they are placed, and so on.

Specify the Quality and the Reliability of the Board. In today's market, the different types of materials have been reduced and higher quality materials are being used more commonly to reduce the manufacturer inventory. Class and reliability issues lay more with the design than the manufacturing of the board. Some military-type boards specify coupons and cross sections of the hole to be done to evaluate the hole wall, along with lot number and board numbering. These options are preformed for tracking and accountability. These issues are decided between the board designer/contractor and their customer. Some specifications that may be used are

- NHB-5300 (NASA)
- Mil-PRF-31032, which specified much of IPC 275
- IPC-6012 (6012 is used for fabrication requirements)

Define: If the boards meet any of these specifications, then this information needs to be documented in the fabrications drawings.

Sample Notes: This board shall meet or exceed the specification of Mil-PRF-31032. Or this board shall meet or exceed the specifications of IPC-6012. (For IPC specs, the actual manufacturing guideline is IPC-6012. It is important to know what values are being specified in any document that is being specified. The designer should read all specifications that are noted.) The class and type should be included in the specification to display clearly the quality of the board requirements. (These specifications are primarily used with IPC standards). The board shall be class II (quality), type III (multilayer).

Specify Tg and Heat. Thermal gradient (Tg) is the temperature value in which a material begins to lose stability or breaks down. The Tg required is determined not by the environment but by the number of manufacturing and assembly processes the board will go through. These qualities are covered further in the Chapter 3, "Design for Assembly."

Define: If IPC material isn't specifically called out, then the required material Tg needs to be specified. This is to ensure that during auto assembly the material does not delaminate.
Sample Note: Not applicable.

Specify the Core Material Type.   Most materials stocked by manufacturers are up to IPC standards. (Check with IPC 4101 for materials.) The thickness tolerance for this material should be no more than +/-.002"[.0508]. The core material is a fiberglass dielectric that is copper clad (Figure 2–1). It is important not to specify the copper or the dielectric thickness at this point. A design may be composed of several combinations of core material separated by pre-impregnated fiberglass, as shown in Figure 2–2.

Note

Dielectric is material that is placed between conductors that acts as an insulator. Core and Pre-Preg are both dielectrics.

Define: IPC material is defined by Tg rating and various other attributes. Check IPC-4101 for the material for the necessary type.
Sample Note: Core Material In Accordance With IPC-4101/21, Laminated Sheet, Copper-Clad, Type GF Glass Cloth Base, Flame Resistant. See Table (Layer Stack-Up) For Recommended Core Thickness, Layer Stack-Up, And Overall Board Thickness (Including Foil And Plating).

Specify the Pre-Preg.   Pre-preg is a fiberglass-type material that is pre-impregnated with an adhesive used to adhere to the core material and to provide an insulator between the copper layers. (Check with IPC 4101 for materials.) The

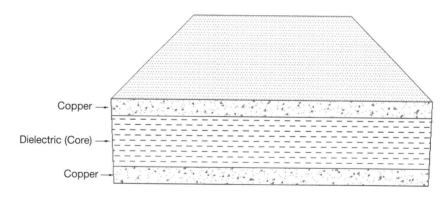

Copper
Dielectric (Core)
Copper

Figure 2–1   Core material.

Dielectric (Pre-Preg) →
Dielectric (Pre-Preg) →
Dielectric (Pre-Preg) →

Figure 2–2   Pre-preg material.

tolerance for this material should be no more than +/-.001″[.0254]. Pre-preg comes in sheets in the thickness of .002″[.0508] each. The pre-preg is stacked to create the desired thickness, therefore the material tolerance is additive.

Figures 2–3 and 2–4 show pre-preg before and after pressing. Pre-preg comes in individual sheets and are stacked together as necessary to produce the desired thickness.

Define: This is the same as the core material. The core and pre-preg material can be incorporated in one single note.

Sample Note: Pre-Preg Material In Accordance With IPC-4101/21; See Table (Layer Stack-Up) For Recommended Core Thickness, Layer Stack-Up, And Overall Board Thickness (Including Foil And Plating).

Note

Do not hold the manufacturer to the minimum core/pre-preg thickness tolerance, unless thickness is critical in cases such as controlled impedance boards.

Define Layer Stack-Up.   There are two types of PCBs: single/double sided boards, and multilayer boards. Materials usually come in a sandwich style, which is perfect for core material. There are separate materials for both core material and multi-layer material because of the way they are measured. Core material is measured by the overall thickness of the material. The multi-layer materials are measured by core dielectric and the copper separately.

As material becomes scarce, more and more manufacturers are using only multi-layer material. When discussing multi-layer material, "1/1–30" or "1 over 1

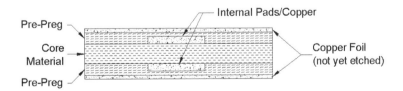

Figure 2–3   Pre-preg material after pressing.

with a 30 core" is used to identify the material. This breaks down to 1 oz copper, sandwiching a .030″[.762mm] core material.

Note

A common difficulty in the PCB industry is the different forms of measurements used. 1 oz is equivalent to .0014″[.0360]. 1 mil is equivalent to .001″[.0254].

Multilayer material also comes in more variations than does core material. Determining which type and thickness to use will be discussed in Chapter 5, "Designing a PCB."

Note

To reduce materials stocked and cost, many manufacturers are using only multi-layer materials.

For multi-layers, the choice is between core stack-up and foil stack-up. Core stack-up is an older-style stack-up that utilizes the core material on the external layers of the board, as shown in Figure 2–4.

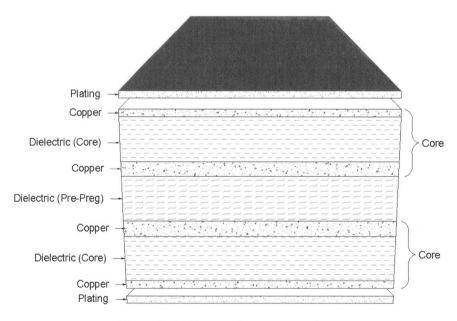

Figure 2–4   Core stack-up cross section.

Foil stack-up (Figure 2–5) is the newer generation stack-up that allows less restriction. For example, with a core stack-up on a four-layer board, two core materials are required and only one layer of pre-preg. This usually forces the manufacturer to take the overall board thickness, subtract the four layers of copper, divide the remainder by three and attempt to use the same thickness material for the two core materials (or a close value), and then make up the remainder with dielectric material. With the foil stack-up, a majority of the board is made up with a core material and the remainder is divided between the two dielectric materials. This allows for maximum flexibility.

There are several combinations of information and values that need to be conveyed to the manufacturer in several different ways. A graphical representation of the board cross section is the clearest fashion and provides a quick and easy reference. This, in combination with specific manufacturing notes, proves to be the chosen combination. Here are the most common pieces of information to provide with the graphical cross section:

- Layer number—Number the copper *sheet* using L1, L2, and so on. Also use type descriptors for the other layers, such as TS for top silk screen, BS for bottom silk screen, TM for top solder mask, and BM for bottom solder mask (Figure 2–6), or a more descriptive table may be used.

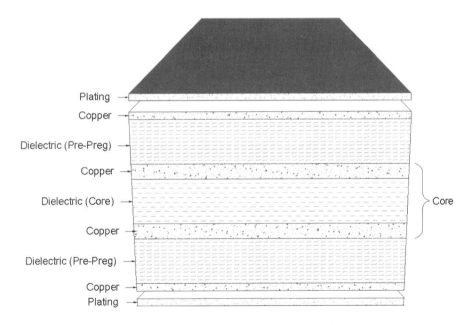

Figure 2–5   Foil multi-layer cross section.

- Layer type—Describe the layer/material type dielectric/core/pre-preg for the dielectric material and for the copper layers, Cu signal/Cu plane/Cu plating. (Cu is the elemental representation for copper.)
- Material weight and/or thickness—A combination of weight and thickness provides a good reference for both the designer and the manufacturer.
- Overall thickness.
- Overall tolerance.

Any other elements necessary that are layer-specific should also be added to the stack-up table, including controlled impedance and minimum trace width by layer. Usually a simple minimum trace note will suffice, but when the trace width values are at the minimum value for the technology, a layer-specific detail may be necessary to avoid confusion.

Define Stack-Up Notes.   Stack-up notes are necessary for a clear description of limitations. The other aspects to note are the material arrangement and some individual tolerances.

Specifying IPC materials, also indicate the material's tolerance. (Check the IPC material specs for those tolerances per material type.)

If the IPC material isn't specified, then the material tolerances need to be specified. Both the core and the Pre-Preg material should specify about +/-2 to 5%. (A .008″[.2032] core will range from .0072″[18288] to .0088″[.2235] and a large .062″[1.574] will range from .0558″[1.417] to .0682″[1.732]). These values are cumulative, and the overall board thickness is the total of all the material tolerances. It is acceptable to specify a smaller board thickness tolerance. All of the materials should not be to the edge of their tolerances.

TABLE 2: MATERIAL STACKUP

LAYER	MATERIAL	THICK	TYPE
TS			
TM			
L1	PLATING	.0014″	
	COPPER	.0007″	TRACE
	DIELECTRIC	.0180″	
L2	COPPER	.0007″	TYPE
	DIELECTRIC	.0200″	
L3	COPPER	.0007″	TYPE
	DIELECTRIC	.0180″	
L4	COPPER	.0007″	TRACE
	PLATING	.0014″	
BM	USE TM		
TOTAL		.062″ (+/−.005)	

Figure 2–6   Stack-up table of the board cross section.

All values should be specified in such a way that the basic generic values are known. This provides a starting point when a tighter tolerance is required. An example is copper thickness tolerance. Normally copper thickness tolerance is not an issue because current tables and formulas have built-in safety factors to account for a small amount of tolerance. Most copper thickness tolerances are +/- .001″[.0254] from the manufacturer, but plating tolerance is more difficult to control. Thus, the designer initially specifies a +/-.002″[.0508] for copper thickness. If controlled impedance type values become necessary, a +/-.001″[.0254] is required. The note for a +/-.002″[.0508] is already given and may then be changed to +/-.001″[.0254]. Notes that are not particularly important act as a placeholder for values when they do become important.

For the stack-up, the designer-specified values are as follows:

- Material types—Specify the material as copper or dielectric. When specifying dielectric, a specific dielectric such as core or Pre-Preg may be specified.
- Copper thickness—Either all layers (finished values) or individually (finished value) and tolerance of each or all.
- Material thickness—This can also be specified individually or overall. Copper thickness may be left up to the manufacturer's discretion and the tolerance of each or all.
- Overall thickness—Specify the overall thickness of all the materials and the overall tolerance. Again, the overall tolerance is a total of all the material. (Many manufacturers recommend a +/- 10% overall thickness tolerance) *Note:* An option is to include an additional thickness such as excluding external copper to specify the designer's intent.

Note

All these values should be specified in finished values.

These values may be displayed in either a manufacturing note or a graphic table. A combination of the two is recommended providing, primarily a graphic description and then a note for those values that are not layer specific.

Define the Material Type.   Specifying either copper or dielectric material is not an option, but the specific dielectric material is an option. If a designer is using common materials thickness, the specific dielectric type may be excluded (Figure 2–6). A note should be used to clarify to the manufacturer that this is the designer's option; otherwise display the specific materials used (Figure 2–7).

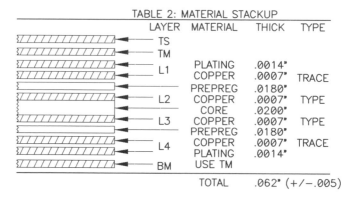

TABLE 2: MATERIAL STACKUP

LAYER	MATERIAL	THICK	TYPE
TS			
TM			
L1	PLATING	.0014"	
	COPPER	.0007"	TRACE
	PREPREG	.0180"	
L2	COPPER	.0007"	TYPE
	CORE	.0200"	
L3	COPPER	.0007"	TYPE
	PREPREG	.0180"	
L4	COPPER	.0007"	TRACE
	PLATING	.0014"	
BM	USE TM		
	TOTAL	.062" (+/−.005)	

Figure 2–7   Specifying the materials.

Note

Use of dielectric material is at the discretion of the manufacturer.

Define the Copper Thickness.   Copper thickness should be specified in the graphic as shown in Figure 2–8 and then the tolerance for all copper thickness placed in a note.

Define: The copper thickness tolerance should be specified if not specified by an IPC specification.

Sample Note: Copper thickness tolerance shall be +/- .002"[.0508] for all layers.

Note

Since copper is in a small range of thickness, this note will usually be on all drawings.

Define the Material Tolerance.   If IPC material is not specified, the material tolerance will need to be noted. Make sure that if IPC material is specified, the material tolerance is *not* specified or verify what the material tolerance is, and ensure that the tolerance specified is the same. Conflicting values will result in delays.

Define: Material tolerances are layer specific or in a +/- value. It should be displayed in the graphic table to explain clearly which layer is which, as shown in Figure 2–9. If the value is common to all dielectric or copper layers, then a single +/- value or % may be used in a note.

Sample Note: Dielectric material tolerance for all layers is +/-.005.

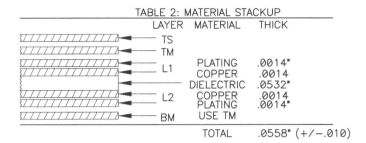

TABLE 2: MATERIAL STACKUP		
LAYER	MATERIAL	THICK
TS		
TM		
L1	PLATING	.0014"
	COPPER	.0014
	DIELECTRIC	.0532"
L2	COPPER	.0014
	PLATING	.0014"
BM	USE TM	
	TOTAL	.0558" (+/−.010)

Figure 2–8   Specifying the copper thickness.

Define the Overall Thickness.   The overall thickness is for the manufacturer's information. If the designer wants to specify a particular thickness requirement, it should be in addition to the overall thickness. The tolerance for conventional manufacturers is +/-10%. This value may be smaller for higher tolerances, technologies, or different materials. The overall tolerance is cumulative of all the material tolerances. In reality, all the materials won't be to the outside of the tolerance, but it is a possibility.

Define: The overall thickness and the tolerance range should be specified in the stack-up table. The overall thickness tolerance is 10% or a +/- range within 10%. Since the material tolerance is cumulative, the less layers or thinner board allows for a smaller tolerance range.
Sample Note: Overall thickness shall be .0XX +/-10%.

Note

This is used only if no stack-up table is displayed.

TABLE 2: MATERIAL STACKUP				
LAYER	MATERIAL	THICK	TYPE	MATERIAL TOLERANCE
TS				
TM				
L1	PLATING	.0014"		
	COPPER	.0007"	TRACE	
	PRE−PREG	.0140"		.0030"
L2	COPPER	.0007"	PLANE	
	CORE	.0280"		.0020"
L3	COPPER	.0007"	TRACE	
	PRE−PREG	.0140"		.0030"
L4	COPPER	.0007"	TRACE	
	PLATING	.0014"		
BM	USE TM			
	TOTAL	.062" (+/−.010)		

Figure 2–9   Material tolerance specified.

Define DS Core and ML Core. Years ago, board materials were measured, including copper. The material was not labeled specifically as double-sided material or core material, although most boards were only double-sided (DS). With the advent of the multilayer board, new thicknesses and types of materials were introduced, and the measurement of these materials was by dielectric and then by copper thickness.

When discussing materials, manufacturers describe these categories as double-sided material and core or multilayer (ML) material.

Fabrication Set-up

Set-up is the portion of the fabrication process in which all the information is gathered, sorted, and defined and a determination of manufacturability is made.

- Manufacturer ability—A CAD/CAM engineer loads the Gerber files, drill files, and fabrication drawings (and notes) and checks for any manufacturing problems or items that are beyond their capability. (This is done to some extent at the quote stage, but a more in-depth check is done at this point.) This also includes materials and material thickness, copper thickness, drill sizes, plating information, and other processes in the fabrication.
- Design problems—This is an optional step that should be confirmed with the manufacturer. The manufacturer may check your design for compliance with specifications made on the design (such as IPC), checking for annular ring and adequate clearances, such as mounting hole clearance and board edge clearances.
- Image set-up—A manufacturer must "grow" the copper areas (for most technologies) to account for etch-off. This will increase all the copper areas, including the trace width and pads; therefore, after etching the copper will be reduced back to its original size.
- Initial routing—The board is still at a "1-up" stage; slots and cutouts are defined and confirmed.
- Panelization—The "1-up" board is panelized, or placed multiple times on a panel (18″[457.2] × 24″[609.6] is common) to speed up the manufacturing process. The panels are placed little more than the router bit width apart.
- Final manufacturing preparations—The drill path for drilling the entire panel is defined, as well as determination of amounts of copper to be placed on or around the board to even the amount of copper on the board. Objects such as coupons or cross-section coupons are placed around the board, and some manufacturer information is placed on the board. If necessary, specify locations and type for all manufacturing markings, such as date code and manufacturer ID.

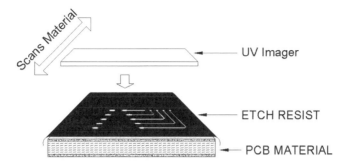

Figure 2–10 Direct imaging.

Imaging

The imaging process transfers the image of a single layer of Gerber information to etch resist (a protective coating) adhered to the copper, by directly imaging the information to the etch resist or by copying from a film that has the image of the Gerber layer drawn on the film.

This is the point where the information from the Gerber files is phototransmitted onto an etch-resist film that is placed on copper-clad material (Figures 2–10 and 2–11).

Define: The view of the data delivered to the manufacturer and how the image is to be transferred to the material (directly or indirectly). This directly correlates to a part of the manufacturer's annular ring. The imaging part of the annular ring for conventional technology is .003″[.0762] using a film (indirect). Direct imaging reduces this amount to between .002″[.0508] and .001″[.0254].

Sample Note: All manufacturers' markings shall be located on the bottom side of board.

Image layer data are viewed from the topside through the board. Images must be transferred to a stable material or directly to the copper material.

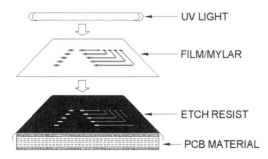

Figure 2–11 Indirect imaging.

Etching

Etching is the process in which a chemical is applied and the copper and unprotected areas are removed (Figure 2–12), leaving the intended circuit. Other emerging processes, such as plasma etch, may be used to accomplish the same task.

Chemical Etch Process

Using the conventional chemical etch process, the unprotected area is etched away (or removed), similar to digging a ditch, in an effective but inefficient process (Figure 2–13). Using the conventional etch process the edge of a trace (and pads) is malformed. For each .001″[.0254] of raise, there is an amount of slope. Trace clearance is measured from the closest point of each perpendicular trace to provide sufficient spacing. This requires more time during etch per ounce of copper to create a larger gap at the opening of the trace, as shown in Figures 2–14 and 2–15. This is known as an etch factor. It is important to know what the manufacturer's etch factor is to calculate the minimum capable per ounce in the event the manufacturer doesn't provide a clear table of minimum clearances per ounce of copper. Table 2–2 shows the amount of width added per ounce of copper to account for the etch factor. Table 2–3 is a quick reference table that shows the minimum amount of clearance, per ounce of copper, to be used to account for the increase in trace width. The "real" minimum clearance is actually smaller than the values quoted, or listed in most tables, capability charts are in reference to 1/2 oz (.0007″[.0178]) copper. This is the designer's minimum clearance, thus avoiding confusion.

The etch factor also affects the manufacturer's annular ring. Conventional AR is .003″[.0762] for imaging + .003″[.0762] for drill + .003″[.0762] for stackup, creating .009″[.2286] overall. Etch or the etch factor is one of the four main controlling items that define the technology.

Conventional etching dictates that the *starting* traces may be no closer than .005″[.127] because of resist fallout, spacing requirements for the etch chemical to work, and so on. In addition, because of the "etch-back" or undercutting that occurs during the etch process, the trace widths must be increased to account for

ETCH RESIST   EXPOSED COPPER

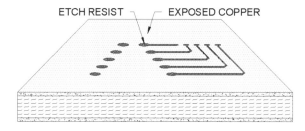

Figure 2–12   Board ready for etching.

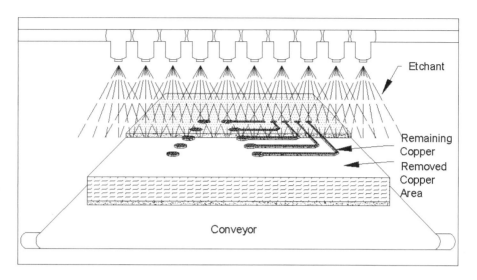

Figure 2–13 Chemical etch process.

the etch-back. This value is determined by the thickness of copper. The thicker the copper, the longer the etch takes to eat away the copper between the traces and under the etch resist.

Two values that must be accounted for in the chemical etch processes are

- Etch factor—The amount of etch-back incurred per ounce of copper.
- Minimum clearance/space width per *starting* ounce.

The conventional etch factor (Table 2–2) is about .0007″[.0118] per size or .0014″[.0356] for the overall trace width per ounce. This means that for each ounce of copper the trace width must be increased by .0014″[.0356].

These values are for the manufacturer to increase the trace width in the artwork/film.

For clearance/spacing requirements, this value is added in addition to the minimum Etch clearance/space, to form the minimum clearance/space per ounce, as listed in Table 2–3.

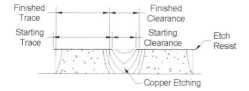

Figure 2–14 Starting and finished trace and space.

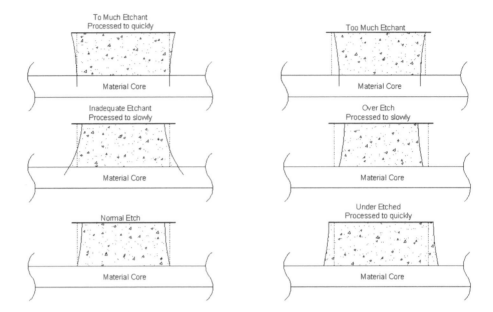

Figure 2–15   Etching errors.

The manufacturer's etch factor (Table 2–2) and the copper thickness determine the manufacturer's minimum trace width (Table 2–3) (per ounce). A 6/6 value usually refers to the minimums or .006″[.1524] Trace/.006″[.1524] space on a 1/2 ounce starting copper.

Plasma/Laser Etch

A new process creating new standards and the demise of the chemical process is plasma etch. In addition to no etch-back, this process also eliminates imaging, or film error using a direct imaging process, which transfers the layer image directly to the material. This eliminates the etch factor, reduces the minimum trace and space, eliminates the imaging error factor, and reduces the manufacturer annular

Table 2–2   Width Added for Etch Factor (conventional)

Starting Ounce	Add
.5	.0007″[.0118]
1	.0014″[.0356]
2	.0028″[.0711]
3	.0042″[.1067]

Table 2–3   Minimum Clearance per ounce of Copper

Starting Ounce	Add	Min. Etch Clearance	= Min Clearance	Round to
.5	.0007″[.0118]	.005″[.127]	.0057″[.1448]	.006″[.1524]
1	.0014″[.0356]	.005″[.127]	.0064″[.1626]	.007″[.1178]
2	.0028″[.071]	.005″[.127]	.0078″[.1981]	.008″[.2032]
3	.0042″[.1067]	.005″[.127]	.0092″[.2337]	.010″[.254]

ring. Compare the conventional MFG. AR (.009″[.2286]) to the leading-edge MFG. AR (.001–.002″[.0254–.0508]). This is a reduction of .007″[.1178], which is the main factor in the reduction of vias and all other nonsoldered plated thru-holes. An .008″[.2032] via, which conventionally requires an .030″[.762] pad, would only require a .012″[.3048] pad using leading-edge manufacturing, reducing the pad by .018″[.4572]. Other methods, such as laser drilling, can reduce the hole diameter resulting in an even smaller pad.

Define Trace Width and Tolerance

Smaller trace widths are more difficult to etch repeatedly; thus manufacturers have a limit to how small a trace they can and will repeatedly process. Because of the etch factor and etch-back, manufacturers will increase the trace width and pad diameters on the image/film so the board will finish at the required width (Figure 2–16). The increase is built into the minimum spacing requirements, and if the clearance requirements are followed, the minimum trace width remains the same regardless of copper thickness.

Warning

Do not attempt to build in thickness for etch-back; this is the manufacturer's job.

A design that has oversized traces, to account for etch-back, will be limited to that technology since different manufacturers use different technologies with different etch-back allowances.

Define: The minimum trace width note is optional but is recommended, so the manufacturer may read the notes and determine the design constraints and cost. As for any values specified, if the tolerance is not specified in the border by tolerance per decimal value, then the manufacturer imposes tolerances. Default manufacturer tolerance (for many manufacturers) for trace and space is .002″[.0508]. This is the recommended tolerance value, but

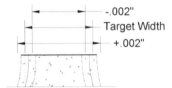

Figure 2–16   Trace width tolerance.

the minimum +/-.001"[.0254] tolerance may be used if required or if the current requirements (in proportion to the trace width) are tight.

Sample Note: Minimum trace width is .006" (+/-.001) [.1524(.0254)]. Min. conductor spacing is .006" (+/-.001") [.1524(.0254)]. Manufacturer shall adjust for manufacturing process
and document all changes.

Multilayer Pressing

Multilayer pressing is the process in which several copper-clad layers are aligned
and adhesive insulate material is placed between the layers (Figure 2–17), which
are pressed under high pressure and heat, forming a solid board (Figure 2–18).
Figure 2–19 displays a board with one layer off center, which is still useable because of the manufacturer's annular ring requirements.

Define: Define the layer-to-layer registration (in reference to layer 1). This provides a constant that all other layers should be referenced from.

Sample Note: Layer-to-layer registration must be no more than .003"[.0762] from layer 1.
For example, "Board scale is not to exceed .001"[.0254] per inch, .005"[.127] overall."

Drilling

Drilling occurs when several boards are stacked and placed side by side and
drilled simultaneously .005"[.127] to .006"[.1524] over the "finished" hole size.
The .005"[.127] (depending on plating requirements, detailed in the "Plating/Hole
Plating" section) over the finished hole size is to account for the amount of plat-

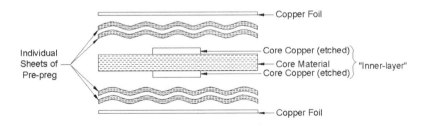

Figure 2–17   Multilayer before pressing.

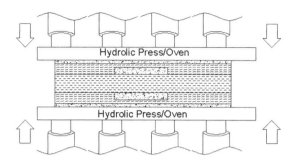

Figure 2–18 Multilayer during pressing.

ing that will go into the hole. Holes that are to be drilled *through* copper but are not plated are drilled at a later time (detailed in the "Second Drill" section).

The manufacturer has limits of how small a drill may be used per board thickness. The rule is, the thinner the board the smaller the drill bit able to be used. This value is the aspect ratio (see Table 2–4). There is the starting aspect ratio and the finished aspect ratio. The starting aspect ratio is the size the hole is actually drilled with, and the finished aspect ratio includes the standard plating. The manufacturer refers to the actual drilled size, and a designer deals with finished hole size. The designer must be very sure what value is being referred to

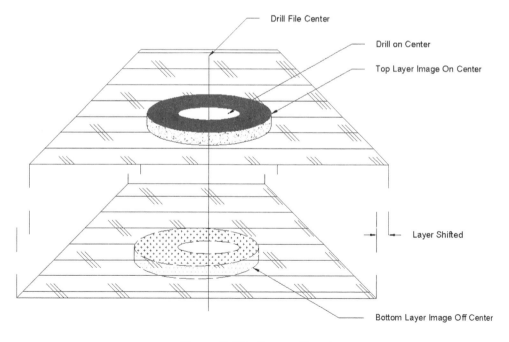

Figure 2–19 Layer off center.

Table 2–4   Quick Table of Finished Aspect Ratios

Board Thickness	For 5 : 1 (conventional) (minimum drill .018″)	For 8 : 1 (advanced) (minimum drill .012″)	For 10 : 1 (leading edge) (minimum drill .008″)
.010	.002	.001	.001
.020	.004	.002	.002
.030	.006	.003	.003
.040	.008	.005	.004
.050	.010	.006	.005
.060	.012	.007	.006
.070	.014	.008	.007
.080	.016	.010	.008
.090	.018	.011	.009
.100	.020	.013	.010
.125	.025	.016	.012
.150	.030	.019	.015
.175	.035	.022	.017

Note: Multiply these values by 25.4 for mm; shaded areas are below minimum drill size.

here. The aspect ratio is limited to the minimum drill, so no matter how small the calculation of the aspect ratio comes out, it may be no smaller than the finished minimum drill. The terms *starting* and *finished* should always be used when discussing drilled size and aspect ratios for clarity. Aspect ratio is also in reference to the board before plating.

Define: Define the finished drill size and the tolerance in a legend, with symbols representing the drilled holes on the board drawing. With the use of the standard ASCII drill file, designers and manufacturers have forgone the symbols, since they add an extra step for comparison. The tolerance (conventional) for holes below .080″[2.032] is .002″[.0508] and above is .003″[.0762].

Sample Note: Drill information in Table 2 shall match drill data file. All drill data is "finished" size. Finished drill locations shall be within .004[.1016]″ of drill data file.

Define: "Hole to drill file registration" specifies the location of the drilled hole in reference to the original drill file.

Sample Note: Hole location shall not exceed .002″[.0508] from drill file locations.

Define: The hole-to-layer-1 registration is the relative point that all other layers are measured to. The hole registration and the layer scale are specified separately and then together to account for SMT boards that have no holes or large boards that have critical mounting holes.

Sample Note: Hole shall not exceed .006″[.1524] from pad center and .009″[.2286] from pad center on all layers (.003″[.0762] for image, .003″[.0762] registration, .003″[.0762] for drill).

Plating/Hole Plating

This is the point at which both the board and the PLTH (plated thru-hole) are plated with copper (Figure 2–20). The boards are placed in an electrically charged bath of copper that plates all of the copper surface areas (normally .0014″[.035] or 1oz) and is drawn into thru-holes located in copper, such as a pad. (This explains why unplated holes that are drilled through copper are drilled later.)

Hole plating is controlled by external copper plating thickness. The hole plating will actually be a little less (roughly .0004″–.0005″) than the external plating. This is critical when determining the current carrying capacity of a plated thru-hole.

As noted in the "Board Finish" section, the board is coated with about .0010″[.025] of solder, making the total thickness of the hole wall .0020 (leaving .0005″ of tolerance). This is a total of .0025″[.0635] on the hole wall and .005″[.127] overall. This forces the drilled hole to be .005″[.127] over the "finished" hole.

The plating or hole-plating note may be skipped if the plating thickness is adequately noted in the material stack-up table.

Boards with blind and buried vias go through this process several times and are treated as several different boards. They are finally pressed together, then drilled and plated.

Figure 2–21 shows the effects of drilling off center to the limits. A drill that is offset should be offset on both sides of the board.

Define: Not all manufacturers plate the same thickness, so for consistency from manufacturer to manufacturer, the plating thickness should be noted. There is no need to specify plating tolerance, since the finished hole tolerance controls the plating thickness tolerance.

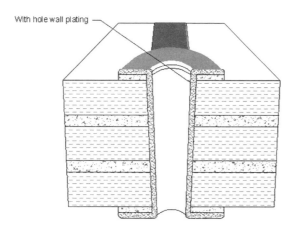

With hole wall plating

Figure 2–20 Hole cross section showing annular ring.

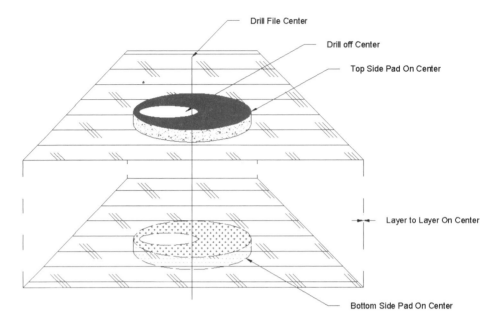

Figure 2–21   Drill off center.

Sample Note (if plating is detailed in material stack-up):
All plated thru-holes shall be plated no less than .0004″ less than the external plating shown in the table titled Material Stack-Up.
Sample Note (if plating is NOT detailed in material stack-up):
All plated thru-holes shall be plated with no less than .0010″[.025] of copper.

Second Drill

Second drill is required when holes are placed in a copper area and the hole should not be plated. Pads around a hole with no plating are known as unsupported pads; they must be larger than pads with plating in the hole for several reasons. The plating keeps the pad from twisting off, dissipates heat, and keeps the pad from lifting from the material when soldered. Such holes add this process, which increases overall cost. If any hole is plated, then all holes should be plated or there should be clear areas around holes in copper areas to eliminate a second drill.

Masking

A protective mask (if specified) is placed on the board for the following reasons (Figure 2–22):

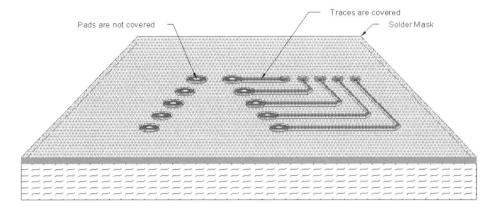

Figure 2–22 Masked board.

- To protect the board from environmental conditions
- To provide insulation
- To block solder bridging between pads
- To protect traces between pads

The mask may be placed on the bare copper (SMOBC recommended) or over a thin film of solder.

Define: The sides that are to be masked, the type of mask used, and the color used. The mask registration should be the same as the mask swell. The mask swell is normally larger than the pad to account for the registration tolerance.

Sample Note: Board shall be solder masked, both sides, using liquid photo-imageable material (LPI). Material shall be green (or other selected color) in color and highly transparent. Registration shall not exceed +/-.003″ Solder mask shall be placed over bare copper.

Optional: Solder mask shall be placed over bare copper (SMOBC).

Board Finish

This is the point in the process after the solder mask has been placed on the board and only copper pads remain. To protect the copper area as well as prepare the area for soldering, the pad is coated with a thin layer of solder. This is usually done with a hot air solder leveler (HASL) (Figure 2–23). It is not necessary to specify the process.

Define: Solder should be applied to all exposed copper surfaces.

Sample Note: All exposed copper areas shall be covered with solder.

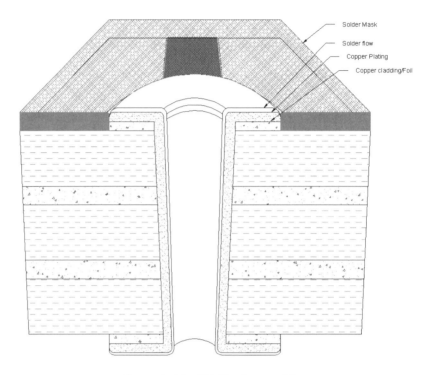

Figure 2–23   HASL after masking.

Silk Screening

At this point the markings for reference designation and component outline are taken from the design and copied to the board.

Define: A good silk screen aspect ratio is 10 : 1 or .100″ high by .010″ thick. The minimum (conventional technology) is .080″ high by .008″ thick.

Sample Note: Silk screen shall be black (or other color) epoxy per A-A-56032. Silk screen to top-layer registration shall be within .003″.

Route

Routing removes multiple boards from a panel while cutting the board to the desired size. Most slots and notches are cut out at this point. Contact your manufacturer about standard holes that can be placed in the boards to help them route the board (tooling holes). All edges and corners are cut with a round bit, so any internal corner must be a radius. If a true 90 is required, contact your manufacturer for the "how-to" information.

The overall board dimension tolerance may be up to +/-.005″. This is the reason behind the clearance necessary on all board edges. Copper-to-edge clearance for board edge, slots, and cutouts is .010″–.020″ + the required voltage isolation clearance.

The outline of the board and all slots should be drawn with a line twice the width of the clearance for "copper to board edge." Note that all cuts shall be made to the center of the line.

Slots that will be plated should be noted and have adequate annular ring (similar to the pad annular ring) along with adequate clearance on all layers/inner layers. It is not desired that copper (especially on inner layers) come in contact with the router bit. (This topic will be explored further in Chapter 5.)

The board outline and the lines drawn for slots/cutouts should be copied to all inner-plane layers to create copper to edge clearance.

Define: The border from the fabrication drawing should provide a guide for the routing of the board. Create an accurate outline of the board, slots, and cutouts. Minimum (internal) radius is .031″ (conventional). If possible, provide three or more tooling holes of .126″ for routing of more difficult boards.

Sample Note: All board edges and cutouts shall be cut to the center of the drawn border. Overall board tolerance shall be +/-.005″[.127]. All slot tolerances shall be +/-.003″[.076] All radius tolerances shall be +/-.003″.

Quality Control

Quality control (QC) is the point at which the board is checked against the designer's specification and the manufacturer's tolerances.

Define: Bow and twist are defined per application. IPC provides a standard to measure for them; .010″ is standard but may be reduced to .008″[.203] if flatness is required.

Sample Note: Maximum bow and twist shall not exceed +/-.010″[.254] per inch.

Thru-Hole Quality Check

When a multilayer board is complete, hole wall quality is a great concern because of the additional factors not found in a SS/DS board. Ask if the manufacturer does a cross section or a cross-section coupon for its multilayer boards. If it does not, request that it be done unless the manufacturer has another process that allows it to inspect the quality of the hole wall that is less invasive.

Electrical Test

Electrical tests may be used to check for continuity and shorts and compare against the original data files. The original data files are the Gerbers, which may or may not match the original design. There are alternatives, such as database transfer formats, that provide a better method of data transfer.

Define: The electrical test may be performed as a lot test (done normally with production runs), where one board is taken from a group, or test 100% of the boards (normal for prototypes). Testing at 100 V is preferred to open any potential line breaks, detect any debris between traces (potential shorts), and find any other type of potential shorting possibilities (i.e., conductive coatings or films). Check with the manufacturer for electrical test capabilities.

Sample Notes: Electrically test all bare boards at 100 V (or minimum spacing voltage isolation) and test impedance of less than 5 Ohm per inch.

SUMMARY

To discuss processes and details of a PCB and the fabrication process, the designer should be familiar with many of the terms used within the industry. Also, there are several levels of fabrication according to the level of technology. These are known as the fabrication technologies. The process for each technology is similar, but the capabilities are different. The manufacturer must know what processes the designer requires and the required values for each process. Starting with material selection and controlling specifications, the designer must specify limits and values for these steps:

1. Set-up
2. Imaging
3. Etching
4. Multilayer pressing (multilayer boards only)
5. Drilling
6. Plating
7. Masking
8. Board finishing
9. Silk screening
10. Route
11. Quality control
12. Electrical test

Following the instructional step-by-step details and using/copying the example notes, the reader should be able to detail the requirements for each step.

3

Design for Assembly

Originally, all PCBs were assembled by hand using only a solder iron. As technology progresses, components get smaller and more difficult to assemble by hand and the amount of components that may fit on a single board increases. Thus the need for auto assembly was developed.

Each assembly process and aspect will be explained in both the manual assembly fashion as well as auto assembly. Soldering techniques are covered in ANSI/J-STD-001.

This chapter deals with the actual assembly process. Considerations for spacing and placement are covered in Chapter 5, "Designing a PCB." This chapter was placed before the design chapter because it is important to understand the constraints of manufacturing and assembly in order to make intelligent, informed design decisions.

SOLDERING A THRU-HOLE COMPONENT

As detailed in Chapter 2, "Design for Manufacturing," a thru-hole component is described as a component leaded with a wire or a metallic lead that is mounted to the board by placing the lead though plated holes in the board and then soldered.

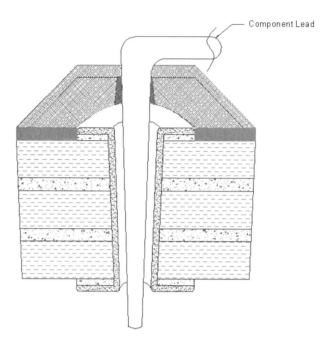

Figure 3–1   Hole and pad to be soldered—step 1.

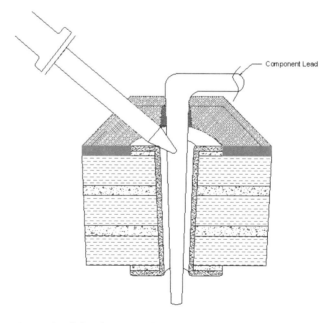

Figure 3–2   Applying heat evenly to the lead and the pad/hole—step 2.

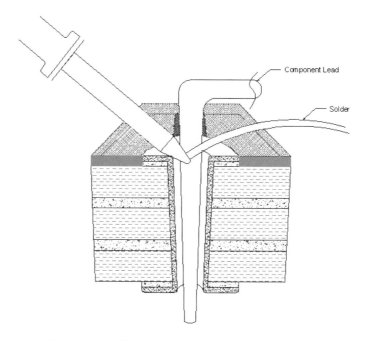

Figure 3–3 Contacting the end of the iron—step 3.

The quality of the solder joint is important for several reasons. The solder joint is the actual connection between the component and the board. The quality of the solder joint is equivalent to the quality of the connection. The "look" of the solder joint is less important but is usually indicative of the quality of the solder joint.

Figures 3–1 through 3–4 display the general steps in soldering a thru-hole component. In step 1, the hole and pad to be soldered are prepared, with the lead being placed into/through the hole. The lead should be placed in such a way as to keep the bend above the soldered area, reducing heat and solder requirements. In step 2, heat is applied evenly to the lead and the pad/hole, heating the material so the solder will adhere to both surfaces. In step 3, solder contacts the end of the iron, which causes the solder to change into a liquid and flow into the hole. In step 4, solder flows through the hole, creating a mound on both the bottom and topside. Since this lead was soldered from the topside, the bottom side should be inspected for adequate solder and joint quality. Some assembly may require solder from both sides to ensure quality. Adequate clearance inside the hole allows good solder flow through the board, allowing only one side to have solder applied. Tight clearances may require soldering on both sides, increasing soldering time.

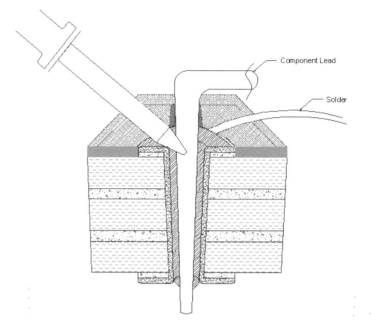

Figure 3–4   Soldering a lead—step 4.

QUALITY SOLDER JOINTS

Conditions such as cold solder joints and solder joints with pinholes and contamination are problems that can be avoided. Quality and expectation of solder joints are covered in IPC-J-001.

Other quality issues are design issues. The designer must work with the solder technician's experience, purchasing component costs, and maintenance's expectation for replacement. Conventionally, fragile, weak, or borderline components on an expensive board should be replaceable to avoid scraping an entire board for simple components.

Figure 3–5 shows a solder joint with an adequate pad diameter. Heat is evenly distributed between the pad and the lead. Figure 3–6 displays a solder joint with an undersized pad. This creates uneven heating between the pad and the lead. The lead has inadequate heating, and the solder joins with the pad and not the lead. Reversing this problem, if a board has a severely oversized pad, too much copper connected to the pad, or no thermal connections between the pad on a plane layer and the plane layer itself, the pad may delaminate from too much heat, as in Figure 3–7.

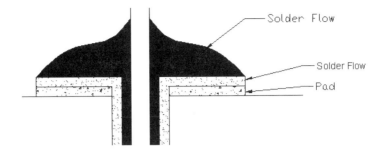

Figure 3–5   Ideal solder joint.

Adequate Pad Surface Area

Heat must be evenly dissipated between the lead, the pad surface area, and the thru-hole. Adequate surface area is essential in the quality of the solder joint. Undersized pad surface area will result in pad delamination (the pad will tear away from the board). The delamination will affect the connectivity of the pad to the hole wall plating. The same applies with the pad diameter that applies to the hole clearance. Refer to the component manufacturer's data sheet for a suggested diameter. If one is not recommended, then these are the questions to ask:

What is the lead diameter?

What is the lead solid?

Is the component itself metallic?

Is there anything connected to the lead that may draw additional heat?

Does the component need protection from heat (i.e., plastic packages)?

The lead diameter is the most common consideration. It should be determined if the lead is solid or hollow. Since a solid lead will absorb more heat and take longer to heat, the pad surface area should be adjusted accordingly.

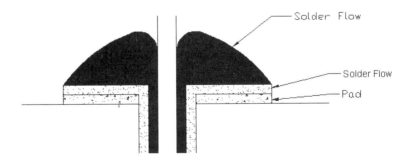

Figure 3–6   Pad to small—poor solder joint.

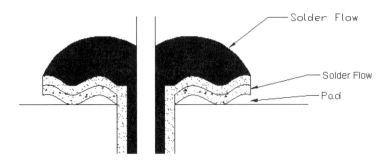

Figure 3–7   No thermals—delamination.

The component itself should be examined to determine heat dissipation/absorption. If a component's lead appears to extend into the component or increases in size (such as relays), then additional heat may be needed to adhere the solder to the lead and the pad surface area should be adjusted accordingly.

Some components, such as plastic ICs, have low heat limits. Some larger items require a heat sink to be placed on the lead between the pad and the body of the component.

Note

In the accompanying software, all these considerations can be used to determine the hole diameter and any adjustments to the pad.

DETERMINING THE ANNULAR RING FOR ASSEMBLY

Component replaceability is determined definitively by the pad diameter and the component to pad ratio. Heat is destructive if not adequately dissipated. If either the lead diameter or the pad diameter is insufficient, or the heat is excessive, one of the items may be damaged. The component lead is the controlling item, and all other factors must follow. The heat must be adequate, but not excessive, or either the pad or the component may be damaged. In the same respect, if the pad is insufficient, the heat balance between the component and the pad will be unbalanced and the adhesive that holds the pad/copper to the board will dissolve and the pad will be released from the board, or may only be damaged, making the board worthless. A pad may be adequate or adequate + overkill. A matrix can determine the size of the pad (or multiplier).

Table 3–1 Expensive Board

Possibility of replacement	Board space requirements	Pad size	
High	Low	Ideal	2 × hole diameter
High	Medium	Ideal	2 × hole diameter
High	High	Nominal	1.75 × hole diameter
Medium	Low	Ideal	2 × hole diameter
Medium	Medium	Ideal	2 × hole diameter
Medium	High	Nominal	1.75 × hole diameter
Low	Low	Ideal	2 × hole diameter
Low	Medium	Nominal	1.75 × hole diameter
Low	High	Minimum	1.5 × hole diameter

This is a component-by-component evaluation and determination of replaceability. Of course, there are many more options to consider, and board expense is sometimes proportional to the space requirements. Therefore, if the pad size is too large, more layers than the expense rating may go from moderate to expensive. These values may be adjusted to accommodate for copper thickness and personal experience.

Tables 3–1 through 3–3 show recommended annular ring requirements depending on the design (board space requirements) and the maintenance (possibility of replacement) aspect. Table 3–1 displays the suggested annular ring requirements for an expensive board, or a board with expensive or hard-to-replace components. The expense determines that the board needs to be serviced as opposed to replaced and discarded.

Table 3–2 Moderate Board

Possibility of replacement	Board space requirements	Pad size	
High	Low	Ideal	2 × hole diameter
High	Medium	Ideal	2 × hole diameter
High	High	Nominal	1.75 × hole diameter
Medium	Low	Ideal	2 × hole diameter
Medium	Medium	Nominal	1.75 × hole diameter
Medium	High	Minimum	1.5 × hole diameter
Low	Low	Nominal	1.75 × hole diameter
Low	Medium	Nominal	1.75 × hole diameter
Low	High	Minimum	1.5 × hole diameter

Table 3–3   Inexpensive Board

Possibility of replacement	Board space requirements	Pad size	
High	Low	Ideal	$2 \times$ hole diameter
High	Medium	Nominal	$1.75 \times$ hole diameter
High	High	Minimum	$1.5 \times$ hole diameter
Medium	Low	Nominal	$1.75 \times$ hole diameter
Medium	Medium	Nominal	$1.75 \times$ hole diameter
Medium	High	Minimum	$1.5 \times$ hole diameter
Low	Low	Nominal	$1.75 \times$ hole diameter
Low	Medium	Minimum	$1.5 \times$ hole diameter
Low	High	Minimum	$1.5 \times$ hole diameter

Table 3–2 displays the recommended annular ring requirements for a moderately priced board. This compromises between solderability and space, thus reducing the pad size and allowing more components on the board.

Table 3–3 displays the recommended annular ring requirements for an inexpensive board, reducing solderability to adequate pad for soldering but not for rework (only one soldering). This provides the greatest amount of room on the board but makes the board essentially a throw-away board.

The annular ring requirements can be determined for the overall board or for individual components on the board. The determination between overall board selection or individual component selection is made by the number of components on the board and the overall cost of the board.

COMPONENT SPACING

Much of the component spacing is for soldering clearance, heat dissipation, and/or noise radiation. Additional considerations will arise with each application, and the possibilities are too numerous to mention. It is advisable to adhere to a standard clearance for all components and increase clearance when necessary. Because most components are soldered from the bottom side of the board, spacing for soldering between components is not much of a consideration. When components are on both sides of the board, the required spacing becomes a factor because of the need for space between components for the soldering iron, as shown in Figure 3–8. When the component's height is below the tip of the iron, space is still of concern, but when components are mounted on both sides it is ad-

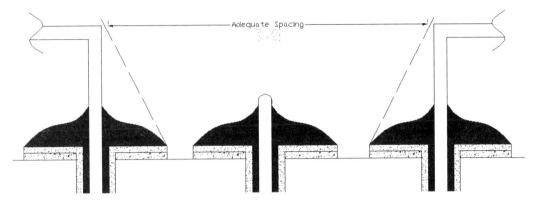

Figure 3–8 Soldering a lead between components.

visable to keep a healthy amount of clearance between the lead and the components around the lead.

COMPONENT PLACEMENT

During component placement, assembly restrictions and needs should be met in addition to physical and electrical requirements. These restrictions or requirements are categorized by assembly type and the cost per type: for example, manual assembly versus auto assembly, and single-sided assembly versus double-sided assembly.

MANUAL ASSEMBLY VERSUS AUTO ASSEMBLY

Manual assembly isn't necessarily less expensive, but the controlling factors are as follows:

    Volume
    Labor expense
    Components used
    Space
    Time

The most restrictive value is space. If space is small enough that surface mount components must be used, then manual assembly becomes more difficult.

When component spacing becomes extremely tight to the point where manual assembly is difficult to impossible, auto assembly becomes a better option. When production quantities are large and time is limited, auto assembly becomes a necessity. Table 3–4 is a matrix used to help determine whether manual or auto assembly is used.

Table 3–4   Manual vs. auto assembly decision matrix

Component types	Spacing/density	Production quantities	Auto assembly	Manual assembly
Thru-hole	High	High	Yes	No
	High	Medium	Yes	No
	High	Low	Yes	No
	Medium	High	Yes	No
	Medium	Medium	Yes	No
	Medium	Low	No	Yes
	Low	High	Yes	No
	Low	Medium	No	Yes
	Low	Low	No	Yes
SMT	High	High	Yes	No
	High	Medium	Yes	No
	High	Low	Yes	No
	Medium	High	Yes	No
	Medium	Medium	Yes	No
	Medium	Low	Yes	No
	Low	High	Yes	No
	Low	Medium	Yes	No
	Low	Low	No	Yes
Mixed	High	High	Yes	No
	High	Medium	Yes	No
	High	Low	No	Yes
	Medium	High	Yes	No
	Medium	Medium	Yes	No
	Medium	Low	No	Yes
	Low	High	Yes	No
	Low	Medium	No	Yes
	Low	Low	No	Yes

SINGLE-SIDED ASSEMBLY VERSUS DOUBLE-SIDED ASSEMBLY

The cost for double-sided manual assembly is not particularly more expensive than single-sided assembly but requires more consideration when designing. (Check with your local assembly house for pricing.) The complexity of assembly support is doubled along with the amount of documentation. Here are some issues that arise when considering double-sided assembly:

> Lead clearance on opposite side (thru-hole components).
> Mounting clearance.
> Additional signal protection.
> Via location. Vias may be placed underneath components, but should be masked.
> Via collision with component on the opposite side.

When an SMT board is designed double sided, blind and buried vias may play a large role, also increasing cost.

Footprints

Few designers keep separate patterns for manual and automatic assembly because of the staggering number of components that would require. The recommended method is to use the larger size (manual) or a little smaller. This may make auto assembly more difficult, but manual assembly requirements cannot be compromised and auto assembly can be adjusted to compensate. If auto assembly is the only method used, then keeping a strict auto-assembly type of footprint library is suggested.

MANUAL ASSEMBLY

There is adequate documentation in this industry on the subject of component lead soldering, so in this section, only those aspects relating to the board design will be covered. A few aspects of manual assembly are different from auto assembly, the most important items being the amount of heat applied and the amount of surface area required.

Some designers maintain separate land patterns: one for manual assembly and another for automatic assembly. Other designers use a generic pattern that works for both applications. The latter is more common with designers who do mixed designs or designs that use both thru-hole and surface mount components.

These types of designs may be soldered by hand if volume is low or soldered by hand for the prototype testing.

Hole Preparation

Hole preparation is normally done in the manufacturing process that plates/or coats the exposed copper areas with solder. This prepares the surface of pads, as well as the hole itself, with a fine coat of solder that will make soldering easier by helping the solder adhere to the pad and hole wall and helping the solder to "wick" through the hole. Figure 3–9 details the hole wall component and pad.

Hole Clearance. Hole clearance is essential for a quality solder joint so the solder will coat the portion of the lead in the hole and allow the solder to flow from one side of the board to the other (Figure 3–10). When heat and solder are applied, the solder will flow around the lead and through the hole to the bottom side of the board, commonly known as "solder wicking." Many solder technicians will solder both sides of the board to ensure even quantity and good quality. Hole clearance is determined by lead size, tolerance, and also lead-to-lead tolerance. The hole should be large enough to allow good wicking of the solder, but not so large that air gaps appear in the solder, creating fissures that may crack when heat/current is applied. Commonly a .010"[.254] clearance is used with most smaller leads or component leads. The clearance does *not* need to be in proportion to the lead diameter because this would result in some very large clearances and poor solder joints. Clearance may range from .008"[.2032] up to .012"[.3048] for smaller components. Larger component lead or mechanical items, such as mounting studs and larger connector leads, may have a clearance up to .020"[.508] The determination of the clearance is more of an application decision. Therein lies the dilemma of components with large lead-to-lead spacing and high spacing tolerance, such as in Figure 3–10. There is no single formula for clearance for a component lead. A complex assortment of

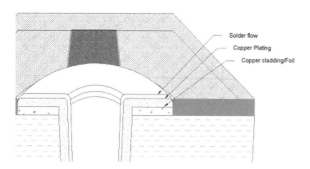

Solder flow
Copper Plating
Copper cladding/Foil

Figure 3–9   Details of the hole wall.

materials, lead shape, in addition to the actual heat dissipated by the body of the component creates an overwhelming group of variables. Refer to the component manufacturer's data sheet or the manufacturer's engineering department. A good component manufacturer should have a suggested hole size for its component. If not, a designer should request that specifications be generated.

If the component manufacturer does not have such documents, then common sense must be applied. Here are questions to ask:

What is the lead diameter?

What is the lead tolerance?

What is the lead-to-lead tolerance?

The diameter itself is the basis of the clearance. If lead tolerance is not specified, use a default of ± 10% or the value may be determined by the convention used for decimal numbers.

$$X.X = 15\%$$
$$X.XX = 10\%$$
$$X.XXX = 5\%$$

The tolerance + value should be added to the lead diameter to provide the worst case value.

The lead-to-lead tolerance is important but should not rule the clearance. Only .001–.002″[.0254–.0508] should be added for larger lead tolerances such as the one shown in Figure 3–11. The soldering technician may need to move or apply pressure to one side of the lead to center the lead to allow solder to flow on all sides of the lead, as shown in Figure 3–12.

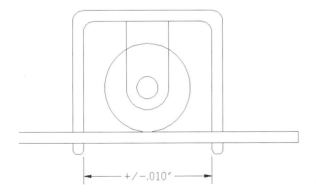

Figure 3–10   Lead in PLTH with adequate spacing.

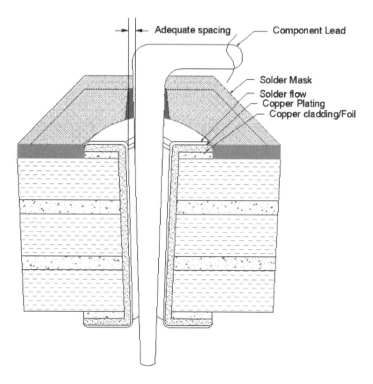

Figure 3–11   Component with a large lead spacing.

Soldering a Surface Mount Component

Soldering surface mount components is difficult and requires specialized tools. Since many surface mount components are sensitive to heat, they are soldered using the least amount of heat possible while maintaining a quality joint. Preparing the surface mount pad with adequate solder becomes important in this process especially since solder is not required to cover the lead but adequately adhere to the bottom and side of the lead or solder area.

The type of soldering used for surface mount components is known as "sweating." This requires the soldering technician to heat the solder enough to make it a liquid and the component/lead sinks into the solder.

Special solder iron tips are made for leaded SM packages. These tips are the length of the package, allowing simultaneous soldering of all the leads on one side of the component.

Leaded packages, such as ICs, are especially critical since the solder tends to "bridge" from one lead to another. This is where solder mask is especially critical. The solder mask between the leads acts as a dam, keeping the solder from bridging across the leads.

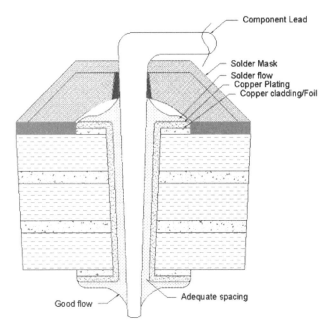

Figure 3–12   Soldered component lead.

   The ability to maintain mask between such small pads is relative to the technology of the manufacturer. The specification is known as the "pitch," or the distance from the center of one pad to the center of an adjacent pad. The actual measurement is from the edge of one pad to the edge of the adjacent pad. The remaining area is then masked.

AUTO ASSEMBLY

Auto assembly is an automated process where most of the component on a board are placed and soldered to the board by machine.

   Auto assembly requires additional information from the designer, in addition to the standard assembly information provided for manual assembly.

When to Auto Assemble

Surface mount boards don't always require auto assembly. Auto assembly is determined by prototyping, volume, complexity, and time.

   Prototyping normally indicates low volume but may be of high complexity. The decision may be split whether to invest money in a board that may not work, or a board that does work and will now have a short turnaround for the production since the set-up has already been completed.

A set-up charge is usually added at the start of a job and is a NRE (non-reoccurring engineering) charge. If a set-up charged for a prototype, it is usually not charged for production. This is sometimes limited due to time and requires production purchase within a week to thirty days.

Even if a board is of low complexity and time is not a large concern, *volume* can become a driving force in determining the need for auto assembly. Manual assembly may be less expensive initially, but when the cost of manual assembly encroaches on the value of auto assembly, then the decision is made.

Complexity is the one "no argument" factor where auto assembly may not be a choice but a must. There are different levels of complexity, such as SMT, tight pitch SMT, or a design with ball grid array (BGA) ICs or sockets. The affordable level of technology for manual assembly must be weighed against the overall cost of auto assembly while taking into account long-term prospects.

Time comes into account in almost all situations. If manual assembly is equipped for the task and there isn't enough time for the auto assembly, then manual assembly is the way to go. If there is a large volume, the time reduced from auto assembly may outweigh the cost.

Auto assembly has been held back greatly by the lack of communication, information, and standardization. Designers are unsure, confused, and frustrated by the lack of standardization from assembly house to assembly house. The following sections provide the information required for auto assembly and the reasons why.

Required Elements

The following are the required elements for auto assembly:

- Gerbers in 274-X (embedded apertures) including fabrication drawing.
- Part centroid text file with Reference Designators, external placement layer, and X & Y location and rotation in ASCII format.
- Numerical control (NC) drill files
- Solder paste file (one of the Gerber files) for all sides mounted
- Glue dot file
- If possible, the design database, specifying the database format (program name)
- Parts list or BOM (bill of material)
- Parts or hardware required

The Gerber files are used to define the pad locations and help the assembly house determine where pin 1 is located and provides a heads-up of how the board looks. The database can also be used to determine the location of pin 1. Some board houses choose to make their own solder paste file/stencil. The designer may create the solder paste file to his/her own preference with experience. It is good to learn how and why the assembly house makes its stencils and duplicate it to ensure consistency from each assembly house (the common name for any company that assembles boards). The solder paste file is used to mask the entire board except those areas that will be soldered. Solder paste is applied to the exposed pads and the stencil is removed. Components are applied and held to the board by the solder paste and the glue dots securing the components as they are soldered to the board.

For component placement a part placement/centroid file is required to know where the center of the part is. The layer ID shows what side the part is placed on, and the rotation displays the orientation of the component.

Consistency in rotation of the original component is critical to report. Unless the designer's software can account for inconsistencies, all components should be created in the same orientation.

NC drill files are used to locate mounting holes and provide the holes sizes for thru-hole components. This also allows the assembly house to determine adequate clearance for the component lead.

A bill of material or parts list is used to reference the designators of the centroid file and the components that need to be mounted. The BOM should also provide information if the component is an SM component or a thru-hole component.

Other Considerations

Other considerations for auto assembly are the board size, the panel size, and breakaways. The boards are commonly assembled in a panel that may contain many boards. The panel is the original material that the boards were etched routed in. The panels pass to the assembly house with all boards intact.

Note

Breakaways are the connections around a board that hold the board during assembly but can be broken easily when it is time to remove the board.

Assembly Limitations

There are several assembly process and assembly lines. An assembly house has limits in complexity or technology, as does a PCB Manufacturer. The assembly house is limited by the following:

The number of parts that can be assembled in one pass

Double-sided assembly

Overall panel/board size

Component size

Component types (surface mount or thru-hole)

Before beginning the set-up for assembly, a survey of several prospective assembly providers should be completed to eliminate undesirable or inadequate providers and document supporting information for those that remain.

The assembly house may not provide the designer or the customer with much information regarding limitations, so the assembly house should be questioned on these points. The most important of these is the number of "passes" (passes is the number of times the board will require the solder flow process) required by the assembly house. These numbers are critical because each pass applies an enormous amount of heat to the board. The Tg (thermal gradient or thermal breakdown; see Chapter 2 for more information) of a board becomes critical. Additional passes require a higher Tg material. A wise rule is to use normal FR4 for manual assembly and higher-temp FR4 for auto assembly. If a board appears to be tightly populated or populated double sided, a higher-temperature material may be used to eliminate any concerns.

Spacing between components or the component's "area" is a consideration during auto assembly. Depending on the assembly house, there are limitations on how close components may be placed (Figure 3–13). A survey of the assembly houses should be taken to see what their limitations are, and these attributes should be defined in the design to ensure that no components are placed too close together.

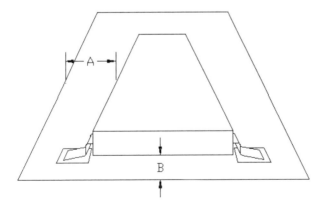

Figure 3–13 IC/component clearance dimensions.

Ordering a Board

If auto assembly is a consideration, an auto assembly quote can be requested before the board is ordered to determine the material requirements. Board material requirements are determined by dwell time and number of passes. These are preliminary findings that must be adjusted to the assembly houses that each designer uses. When dwell time and passes are determined, they should be documented in the assembly notes or the material chosen should be able to accommodate the worst-case scenario.

Once a set of rules is in place, it is a good idea to obtain quotes of assembly costs before manufacturing. Doing so will help spot troubles before manufacturing begins and eliminate rework of the raw board.

SUMMARY

This chapter explained the assembly process, specifically manual assembly. This chapter explains the reasoning behind the area necessary for surface mount and thru-hole pads, specifically soldering quality. Soldered pads should meet manufacturing requirements, but the soldering requirements usually require even more area. The necessary pad area isn't exacting but does have minimum requirements. There are several issues that determine the pad size and lead spacing:

Component cost
Area required
Overall board cost
Amount of boards in production
Overall cost
Overall time

All of these issues must be dealt with *before* the board design starts, allowing the designer to select/create pad styles and to determine if there is adequate spacing in the area allowed. Some of these decisions may be made after the initial design starts and new issues come to light.

4

Schematics
and the Netlist

This chapter deals with the layout of a schematic and the intelligence behind a schematic, or the Netlist, and attributes. A netlist can be one or all of the following: a point-to-point list, a list contained within the program, or a text document, such as P1-Pin1 to P2-Pin3. An attribute is a description or characteristic. This may be a value, description, or title. With schematic and PC boards an attribute refers to a value attached to a component, design, net, or any item in the design. Additionally, symbols and standards for components as well as component creation will be covered in this chapter.

SCHEMATIC ENTRY

As boards become more complex and the implementation schematic simulator tools become more prevalent, circuit designers/engineers will lay out their own schematics more often. As functionality becomes easier and streamlined, circuit designers will steer away from dumb schematic drawings for the schematic capture programs and eliminate the redundancy of schematic reentry. Until then, some board designers will have to decipher what the circuit designer's intentions are and find that the inconsistencies of the circuit designer aren't desirable to duplicate (e.g., gates/sections of ICs).

Schematics are driven by serviceability and replaceability of boards. If this isn't a factor yet, it should be. A service technician depends on the schematic and on a clearly labeled board to determine problems with the equipment that houses the finished product. The following sections will help determine the theme of the schematics and ultimately the PCB.

First, some of the essentials of the schematic must be understood.

UNDERSTANDING ELECTRICITY

Without extensive discussion of electrons, a circuit is a combination of a positive (+) line and a grounding or negative (–) side. Electricity is comparative to a vacuum cleaner. A positive line, by whatever means, has less electrons than ground. When this event occurs, a vacuum effect is created, and a rush of electrons from ground to the positive side is electricity. A difference between the positive charged item and ground (potential) combined with resistance determines the voltage. The amount of electrons flowing is current. Voltage is relative to pressure (or suction in our vacuum example), and the amount of airflow is current.

As in a vacuum when the pressure/vacuum is increased the airflow/current is increased likewise.

If the pipe is pinched or reduced, the airflow is restricted and decreases the flow/current. This is relative to a resistor. With the basic understanding of electricity all of the following is easier to understand.

SOFTWARE TERMINOLOGY

There are some terms that are used in schematic software that have a little different meaning outside of the industry. Some terms are slang and their meaning depends upon different processes being discussed. The following terms are not specific to PCB design software but are used in the context of schematic entry.

Footprint—The PCB representation of the pads, outline, and mounting holes for a component.

Symbols—The schematic representation of a component, displaying important input or output functions.

Block symbol—A symbol representing the entire package.

Gate symbol—A symbol representing on one specific portion of a component.

Heterogeneous gate—Gates of different types.

Homogenous gate—Gates of the same type.

Pins—The symbolic representation of the physical pins or wires of the component.

Wire—Traces that connect parts electrically and will be transferred to a PCB. Wires are usually intelligent items or have attributes and definitions of components they are connected to.

Line—A dumb line representing used to represent external board items, or any items with no electrical connections.

Page/sheet connector (Figure 4–1)—A sheet connector or page connector are essentially the same object but are sometimes handled differently. They may be a reference from one page or sheet to another, representing either one single net or a bus.

Bus (Figure 4–2)—A bundle of nets, represented by a single (wider) line to reduce the number of lines displayed.

Attribute—An attribute is a widely used term but is essentially any information pertaining to any element in a design. The entire design may have attributes, such as size and number of sheets. A net has attributes, such as current and voltage. A connector has attributes, such as clearance required, height, size, and symbol type.

IEEE symbol—Symbols that are consistent with standards set by IEEE.

De Morgan symbol—Block symbols with characters, symbols, and acronyms on each pin to represent pin function and type.

Grouping block—Uses lines and dashes to group connectors or sections of a board. They may overlap from one column or page to the next. The beginning usually starts with a flat line, and each break in the block ends with a continuation line.

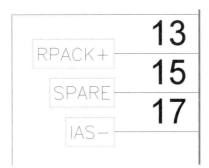

Figure 4–1   Sheet connector.

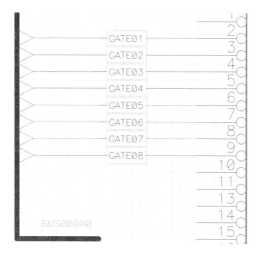

Figure 4–2 Bus connections.

Continuation line (~)—Used to break a grouping block and also used to show the continuation point at the beginning of the next related continuation block.

Net—A connection or group of points that are commonly connected. These connections have the same voltage but not always the same current, as in Figure 4–3. The +5 V line is connected to P5 Pin 35 and P4 Pins B17 and 13. This is a single 5-V net. In this specific schematic, the current flows from P4 Pin 13, to P5 Pin 35, in this case 2 Amps. Current flow from P4 Pin 13 to P4 Pin B17 is only .025 Amps.

Class—A grouping of nets similar to any class system. Nets are grouped by common attributes. Normally, a range of voltages and currents or layers will group nets. In Figure 4–4 the +12 V and the +15 V aren't too different in voltage, and spacing will not be much different either. The +5 V and +24 V are substantially different and will be placed in their own classes. When defining the attributes of the class, the worst-case value is used. In this case the class, named "Power 1," will contain the +12 V and the +15 V nets and the attributes will be Amps—1, Volts—15. This defines the worst-case values. The values will then be carried over to the PCB layout program to be "translated" into width (trace) and clearance (space). Physical PCB attributes should not be defined in the schematic portion since board attributes, which are not known at this time, are required to determine the width and clearance of the board. The determination of grouping is defined by the amount of room available on the board. If space isn't a concern, all of these nets could be grouped and the attributes would be

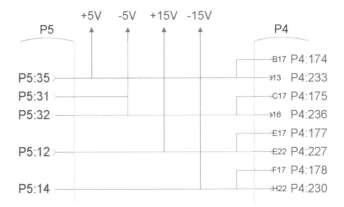

Figure 4–3   Net connections.

Amps—2, Volts—24. This would cause all widths and clearance to be the same.

Class-to-Class—A way to define the difference between groups. For example, upper class and middle class may be defined by the difference in the amount of money, so the net classes are defined by the difference

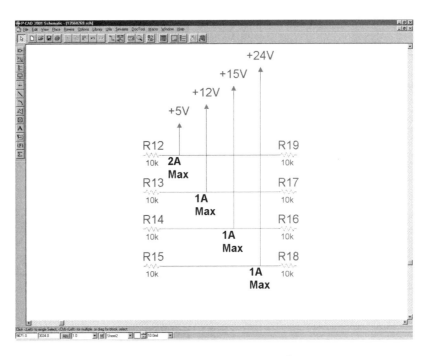

Figure 4–4   Nets to group by attributes.

in voltage or the potential. The potentials are defined in reference to a common point or ground point. In this example the common points for all are the same, and there are three classes:

Power 1—Contains the +12 V and +15 V nets
Power 2—Contains the +5 V net
Power 3—Contains the +24 V net

All combinations must be defined to ensure adequate clearance.

Power 1 to Power 2—The potential is 10 V (worst case).
Power 1 to Power 3—The potential is 9 V (worst case).
Power 2 to Power 3—The potential is 19 V (worst case).

Unlike class voltage/clearance attributes, whose potentials are from similar voltage, class to class deals with dissimilar voltages, making it necessary to define class-to-class attributes.

A +30 V class and a –30 V class have only a 30 V potential within the class, but from class to class there is a 60 V potential. The spacing would be twice that defined in the ± 30 V class. Different currents are not necessary to define since they have no bearing on class-to-class definitions.

Room/Area—An area on the board, on one or more layers, that has specific attributes such as clearance, voltage, noise, or component restriction.

Other Attribute Definitions

Items such as a board or layer attributes are defined similarly but are specific to the entire board or to a specific layer. If a board's width attribute is .012″[.0348] and a layer width attribute is set to .006″[.1524], the layer attribute will take priority. This is because of the common hierarchy of definitions. Here is the list from most important to least important:

Connector pin—The actual metallic contact point(s) of a connector.

Female connector—A connector that is recessed in a sleeve or a connector that a mating connector can be placed into.

Male connector—A connector that protrudes outward and will fit inside the mating connector.

Female pin—The pin that is recessed and the mating pin may fit inside of.

Male pin—The pin that protrudes outward and fits inside the mating pin.

Power pins—Any pins that will be used solely for input power to an IC.

UNDERSTANDING COMPONENTS

There may be many mechanical objects on a PCB that may be stored as a component for ease of use, but we will reference a component as any object in a schematic that is electrically connected or requires placement in the design. It is understood that many components have a positive and negative connection. This is not always the case, but in most situations it is. In addition to a positive and negative connection, a component may have additional lines either defined as inputs or outputs. Inputs affect the component or bring about an effect, and outputs are usually the result of the input as shown in this simple gate (Figure 4–5).

Not shown in Figure 4–5 are the positive and negative connections. This is quite common, and positive/negative connections should be defined elsewhere on the schematic but are eliminated in this symbol to reduce size and area used. Pins not shown or "invisible pins" are a common part of schematic but must be shown in some area of the schematic, whether a table or in text. (In Figure 4–5 pins 1 and 2 are the input and 3 is the output.)

Symbol Types

There are several formats of symbols that may be used, but the two common groups are the block style and gate style. Figure 4–6 is a block style, which is a symbol representing the complete component.

The other type or "gate" (shown in Figure 4–5) represents only a portion of the component, and in some instances the gates may be swapped to use the best-positioned pins for smoother routing.

Components Display

There are several ways to display components as well. The intricate options of a component are covered in Chapter 6, but the overall standards are covered in this chapter.

The schematic software usually has components that comply with many of today's standards, such as IEEE and De Morgan symbols. IEEE is the more common but less informative, whereas De Morgan displays the functions of the pins and the way they are triggered.

7408   **Figure 4–5**   A simple gate.

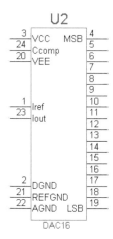

Figure 4–6 Block style.

Schematics should utilize only one style for consistency, but some may require IEEE and the occasional De Morgan for a component that is very important. Check into each style and determine the pros and cons for each style. The two main styles of component display are in block form or in gate form.

Logic ICs and relays are commonly broken into gates or individual symbols representing parts of a whole. Some designers break block-style components into separate sections, sometimes grouping them by input and output, digital and analog, or by function. These are all acceptable practices and may not always be consistent throughout all designs, but only when the schematic or the design engineer requires it. This is especially common in components with a numerous amount of pins. Displaying the component as a whole may become too complex and too confusing. The component may be of such a size that only the component itself and sheet connectors are displayed. This is undesirable to many service technicians but practical and necessary. Some components may be so large or have so many pins that the component is broken into sections and displayed on several pages as though it has several gates.

Component pin numbers and names are also a consideration for troubleshooting ease. If technicians are required to troubleshoot down to component level, then pin names are necessary on the schematic. Pin numbers are almost always necessary, not just for troubleshooting purposes but for layout purposes.

Net Names

Net names seem a simple determination and are often automatically numbered or named. The need to name a net is largely dependent on the availability of room and the requirements of troubleshooting. Net names are often used to ease display or show the function or origination of a particular net. An example is a multi-board assembly where a net may pass through several boards before coming to its destination. A simple name, such as the voltage, type logic level, output pin, or function, may be necessary, as follows:

> Voltage. (+5 V, +15 V, V+, +Logic, GND, Digital GND, Analog GND, Chassis GND)
>
> Type (clock, enable, disable, strobe)
>
> Logic level (DO1 High)
>
> Output pin (DO1, DO2, Out1, Y1, Z1)
>
> Function (similar to type)

SCHEMATIC STANDARDS

Often designers will have to either enter a schematic or define values necessary for the board design (a feature used in most schematic capture programs).

A PCB design is not always started with a schematic, but often this is the case. Traditionally the PCB is a compilation of schematic entry by an electronics engineer (or the equivalent) and a PCB designer. As technology progresses and PCB boards become more complex, one person may do the entire job. A designer should be able to understand a schematic, determine situations, and understand the needs of the engineer. This flexibility in a designer is a good selling point in today's world.

Sheet sizes, drawing standards (see Table 4–1), and output format are a product of a company's requirements, such as the following:

> *Publishing*—The format of the manual or technical manual requires certain paper size. Some are small with pullouts or letter sized with ledger-sized pullouts.
>
> *Government*—Some government agencies or contractors require all drawing documentation to be on a D size drawing.
>
> *Company Standards*—Many companies may commonly use a B size drawing that uses less page connectors than an A size (letter) but is more manageable than a D size drawing. This is the recommended size, allowing printing on A size to D size paper.

IEEE defines most standards in schematics and electronic symbols. Many of the symbols and reference designations follow some conventions but often follow personal or company standards. Table 4–2 shows some of the common letter(s) used to represent components.

Table 4–1 Schematic Style Standards

	Yes	No
Sheet size A used?		X
Sheet size B used?	X	
Sheet size C used?		X
Sheet size D used?	X	
IPC drawing standards used?	X	
IEEE standards used?	X	
ANSI drawing standards used?	X	
Sheets in Columnar format?	X	
Sheet in generic end-to-end format?	X	
Sheet in continuous sheet format?	X	
Sheet zones shown?	X	
Continuous sheet zones shown?		X
Sheet connectors show page and zone?	X	
Sheet connectors show zone only		X
Sheet connectors show net name only		X
Connector block?	X	
Connectors individual?		X
Connector shows block/grouping lines?		
Components shown as blocks?	X	
Components shown as De Morgan?		
Components shown as IEEE?		
Components shown as gates?	X	
Filter caps shown in block with component?		
Components display component type?		
Component pin names shown?		
Power pins displayed?		
Power pins hidden? (Use PP table.)	X	
All net names displayed?	X	
Selective net names displayed?		
No net names displayed?		

Table 4–2   Component Letter Standards

Letter	Component	Letter	Component
K	Relay	Z	Zener diode
R	Resistor	W	Wire
J or P	Connector	X	Transformer
D	Diode	Q	Transistor
C	Capacitor	F	Fuse
L or H	Inductor or Coil	R	Potentiometer
L	Led or Lamp		
T or TP	Test point		
Symbol only	Ground (common)		
Symbol only	Ground (Chassis)		
Symbol only	Ground (Digital)		
Symbol only	Ground		

Of the most important requirements, consistency and standards top the list. All aspects of design are easier when values are defined and documented. Standards such as grid size control the entire drawing as well as increase the ease of editing and wiring. All the symbols should be designed on these grids to allow easy modification and pin spacing. Some components, such as gates, have a single pin on one side, making it difficult to center. The grid may be reduced to the lower values to accommodate this. The scheme in Table 4–3 allows the schematic to be wired, using the .200″[5.08] grid, and reduce to .100″[2.54] for smaller components or off grid pins.

Table 4–3   Schematic Control and Text Standards for Primary Drawing Format

Drawing sizes	A&B	C&D
Components pins	.200″[5.08]/ .100″[2.54]/ .025″[.635]	.400″[10.16]/ .200″[5.08]/ .100″[2.54]
Text heights		
Reference designator	.100″[2.54]	.200″[5.08]
Connector reference designator	.150″[3.81]	.250″[6.35]
Component values	.075″[1.905]	.150″[3.81]
Notes	.100″[2.54]	.200″[5.08]

Note: These values depend on the program's text measurements and may need to be adjusted to fit other formats.

Note

A fixed set of grid values should be used. If a metric system is used, a different set of standards and measurements should be used (e.g., 5, 10, 25, 50, 75, 100 mil).

The values in Table 4–3 allow for consistency and separation of text type by size. The sizes for connectors are larger, since they are commonly searched for in a schematic. The text sizes for the C&D size drawing allow printing of D size drawings on a B size paper if necessary. If this is not desired, the same size should be used throughout to support only one single text size. Some customers may request all schematic drawings on D size paper. This is the only time that using these sizes is suggested.

Some programs allow naming conventions of the text styles. This is a very convenient tool in standardizing. The user can name the style by the commonly used "type" of text, such as:

RefDesg—Reference designators

RefDesgCon—Reference designator connector

PN—The text placed on the board showing the part number

Pin number—Numbers for pin location

Title—Title text

Zone—Text that represents the zones of a drawing

If all aspects of the design are based on these grid standards, all items become easier to edit and move. Designs look professional because of consistent spacing between components, connectors, and wires.

SCHEMATIC DESIGN CHECKLIST

Circuit design requirements are the first stage when initially planning any design. This is where either the designer, engineer, or both discuss and decide by what method the schematic should be laid out and decide the size of the board(s) and if the board size can contain the entire circuit or should be divided between more than one board. The decisions may be made based on conceptual drawings or an idea of what is required. This is event and application driven. The age-old debate of form over function or function over form is decided. Neither is correct, but it is decided by application. A board size and space requirements may be predefined and the schematic is divided across more than one board, or the area and board size may not be very important and the size is determined by the schematic size and need.

SCHEMATIC STYLES

The schematic style is determined in large part by serviceability. If the board being designed will be serviceable or may need troubleshooting, then the schematic must be handled differently. A wise way to begin is to decide that all of the boards designed will require service and start from there. Why is servicing a concern? When a technician is attempting to find a part on a board, there should be a logical flow to the schematic. Components should be easily found, grouped, and laid out in somewhat the same manner as the board. This is not always easy or necessary but a good practice. Here are some of the different styles of schematics:

- Logical flow is determined by the type of circuit and flows through the pages similar to the circuit. This may also be mixed with some of the other styles.
- Smaller boards or simple boards may be laid out similar to the intended layout of the board. Some boards have a predefined layout because of the connector placement requirements.
- Connector layout is similar to board layout style but is less restrictive and more common and only follows the location of the connectors. This allows for similarity of connector placement while maintaining a more logical flow.
- By function may encompass either or both of the previous two styles or follow only function over form (if room allows for more than one function type per page, but the attempt is to break each page into a function type or power type, as noted in "Sheets and Strategies"). This style is the most common for multifunction circuits and/or very large boards that contain items such as control circuits (using digital control or low voltage signals to control/switching relays for power control) and is the best for documentation and use with an index by function.

SHEETS AND STRATEGIES

The media in which schematics will be displayed take on a roll in the strategy of the schematic layout. If the schematic is used in a small service manual, then special steps may need to be taken for the layout. General documentation may dictate that either an A size sheet or a B size sheet (in fold-out style) is used. Some customers and designers require all documentation be in D size drawings. This is the first step in determining the way the schematics are done.

Naming the sheets will come toward the end of the layout but must be considered, and a target should be devised before needless moving of circuits occurs. Some of the common sheet names/labels are as follows:

Power supply
Input power
Filtering
Logic
Power
Power switching
Contact section
Coil section
Metering
Sensors

This style (naming the sheet by function) provides a quick and easy reference of what the page is about and what should be found on that page.

The border of the page is a secondary concern. If following electronic standards of documentation, then zones and title blocks will be used. It is recommended that IPC or some other organizational standards be used. Many layout software packages will have predefined borders and title blocks for your convenience. When implementing these, take into account areas for revision, sheet revisions, title blocks, and notes. Use the worst-case scenario for long-term design practices and templates.

Templates also provide consistency and structure. A schematic may be displayed in a columnar form (dictated by sheet size and personal preference) or in an end-to-end form. Columnar form provides the most consistency but also the most restriction. Column lines may be placed along with text and connector start–continuation–finish points.

CONNECTORS AND SHEET CONNECTORS

Connectors are the focal point of many designers because of the mechanical requirements and their required placement. This poses the first schematic question that utilizes the design philosophy: How should this connector be displayed—as

Figure 4–7   Entire connector in one location.

one piece (Figure 4–7) or located near their connections (Figure 4–8)? Placing the
connection near the component it will be connected to will reduce the size of the
schematic but causes some confusion as to where all the parts of the connector
are, and each part/gate requires individual labeling. When connector location is
critical and reduction of schematic size is crucial, a "bus" may be used to connect
connectors and other pages or components. A bus is a single (but thicker) line that
behaves as a bundle of wires for simplicity.

Sometimes the number of connections is so numerous for one connector that
grouping all the pins of the connector is not an option. When grouping the pins is
an option, often a sheet connector (Figure 4–9) or find number is used to connect
the two or more connections whether they are on different areas of the same page
or on different pages. If zones are used on your schematic, then they should be
used with the sheet connector to locate the corresponding connection. Some
schematics use continuous zone numbers (the number at the top of the page)
across each page of the schematic to eliminate the sheet number and use only the
zone number.

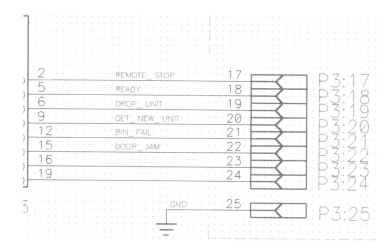

Figure 4–8   Connector placed separately.

If a continuous zone format is used, only the zone name need be displayed, such as 3B, 5A. The zone direction is determined by the format of the schematic. If the schematic is drawn on a horizontal plane, the vertical zones will remain the same throughout the schematic. If the schematic is drawn in a columnar format (vertical), the horizontal zones will remain the same throughout the drawing.

If possible, a connector should be displayed in its entirety, such as Figure 4–10, reducing time and confusion of locating the rest of the connector. If the connectors are shown as one continuing connector but the number of pins exceeds the height of the paper, special symbols are used at the end of the first and beginning of the next connector block (Figure 4–11).

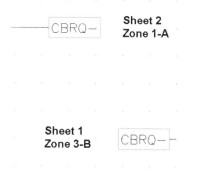

Figure 4–9   Corresponding sheet connectors.

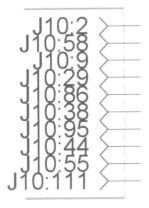

Figure 4–10   Regular connector block.

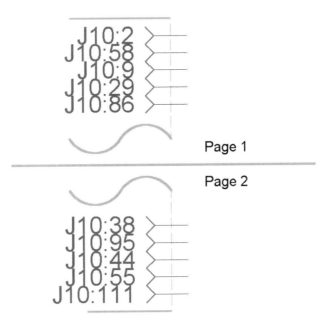

Figure 4–11   Continuation of a connector.

Designer Checklist Alert!

To better understand connectors and sheet connectors, the following discussion incorporates portions of the Designer's Checklist into appropriate sections in a quick overview format, followed by more detailed explanations. When a checklist doesn't apply, stand-alone discussions are presented.

A break line shows that the connector is not shown in its entirety. For easy reference, text may be used to show where the other portion(s) of the connector are for easy reference. These are all standards that must be documented to provide consistency.

❏ *Gather required information for part list, required component locations, and mechanical locations and requirements.*
❏ *Determine if all components are available in existing libraries. If not, use a component creation checklist.*
❏ *Select design template.*

This may include required sheet size, border styles, and templates with predetermined/commonly used parts.

❏ *Save file by part number.*

Saving the schematic by a part number allows quick referencing, searching in relation to the PCB, and conversing with the board manufacturer.

❏ *Enter design information.*

This is dependent on the way your program is set up and/or if information in the schematic files require manual text change/entry.

❏ *Open/load necessary libraries.*

Check the parts list and note on it the library names used, and/or set up the libraries to be used that contain the necessary components. If some components don't exist, create them using a component creation checklist. It is good practice to create the footprint before attaching the schematic symbol because of mechanical requirements of the footprint, such as mounting holes and optional mounting positions. Some pin symbols may be overlaid to appear as one pin to provide the options of different footprints or different mounting positions.

❑ *Place components and wire together.*

As noted previously, component placement and wires should be placed on the grid and an attempt should be made to keep them in some standard orientation. Sheet connectors should be placed in opposite directions from the initial page to the continuing page to show connectivity. Even horizontal or vertical locations may be followed to allow an overlay of the sheets to provide clarity of connection.

❑ *Note all current, voltage, high frequency, noisy, and sensitive circuits.*

For high current circuits, note current of branch circuits.

Defining branch circuits and their current requirements is important if the main branch is of very high current. If the main branch requires a .050″[1.27] trace and the branches only require a .008″[.2032] width, it makes no sense to route all branches with a .050″[1.27] trace.

❑ *Add a note like the following for the most commonly used trace and space in the board designs: "Unless otherwise specified, all circuits are less than .25 A and 30 V" (this works for 6/6 {.006″[.1524] trace and .006″[.1524] space}).*

This note should work for the most common of traces to reduce notation of voltage and current. Logic or digital designers that use only conventional technology require the 6/6-type note. Analog designers will typically require a note that correlates with 12/12 trace/space.

❑ *Place power-pin table.*

A power-pin table is used only in the event that the power connections are not shown on the symbol. Because of the space required for + voltages and ground symbols, most software packages allow there to be hidden or invisible power pins, allowing a connection to a predefined value or connection (net). These pins are not displayed, so the option of a table displaying the RefDesg and the pin number is placed in a column of the relative net name or value. This takes up a small amount of room and allows quick checking of the power connections.

❑ *Place "last used" and "unused pin/gate" table.*

If an option, a "last used" table may be placed and later removed. This provides an accounting of all the last RefDesg used and any skipped, allowing for a successive count of components. The "unused pin/gate" table provides a list of all unused pins in a part/gate that has been placed. This is useful in multiple symbol components and helps identify or troubleshoot pins that have been left out.

❑ *Highlight power nets and check each sheet for connectivity.*

Highlighting power pins ensures that all connections to sheet/power pin connectors are connected correctly. Misspellings or mislabeling of power connections is a common cause of unconnected components. A simple inspection of these, and any net using sheet connectors, can save hundreds in reworked board, time, and labor.

❑ *Check for design rules, such as single node net, no node net, unconnected pins, or unconnected wires.*

A design rule check, better known as a DRC, can be conducted at any time to inspect several options of the schematic and components. Options such as outputs of a component connected to other outputs may be caught early in the design using these options. Many prefer to perform a DRC at the end of the schematic layout as a final electrical check.

❑ *Generate a BOM and compare against a parts list.*

This is an option, but it is a recommended process to ensure that the correct components were used.

❑ *Add necessary notes.*

Notes can be used to make generalized statements about values, wattages, tolerances, or functions. For example, "All resistors are 1/4W, 2% unless otherwise specified." This reduces the number of values that must be displayed on the schematic, reducing the schematic size and overlapping of values and circuits.

❑ *Add sheet numbers.*
❑ *Print and check schematic visually.*
❑ *Align/modify location, format, and styles.*
❑ *Place all nets into classes.*

Define all classes with the minimum of current and voltage attributes.

❑ *Generate netlist.*
❑ *Archive libraries.*
❑ *Check for other concerns:*
    ❑ *All IC inputs terminated as required?*
    ❑ *Do IC/components have necessary filter caps?*
    ❑ *Are main circuit and branch circuits clearly identified?*

Note

Many of these functions depend on the design software. Adjust the steps and syntax for personal use. Define as detailed as possible those functions or settings that are always followed.

SUMMARY

There are several different styles and formats by which to display a schematic. There are standards that may be followed that will reduce confusion and increase consistency. The choice of styles and displays depends on the application. Following standards and conventions is essential. Many are provided for your convenience. Following and repeating a proven method will reduce errors and provide consistency throughout designs and provide consistency and clarity for the PCB designer.

5

Designing a PCB

The information in the previous chapters set the stage for what you are about to learn here, where the board is actually designed. This chapter details the materials, thickness, and manufacturer capabilities and combines that with the design requirements, including available area and mounting styles, and utilizes the accompanying Designer's Checklist, providing a complete resource for board design. This chapter was written to instruct the design process, including basic design philosophies, routing methods, and stack-up styles, while supporting the embedded Designer's Checklist.

Note

The Designer's Checklist is a comprehensive detail of the design process that may be used with every design and customized per the design's specific application. This checklist provides a proven structure while creating consistency from design to design. The complete Designer's Checklist is provided both in Word format and as a PDF file on the CD, and in the book's introduction.

INITIAL DESIGN DETERMINATION

Parts of the design process are repeated in this chapter because an initial estimate of the design requirements and technology used is made according to several constraints. This is to provide an estimate of the board design and provide a guideline for the design. During the design, a decision may be made to exceed those initial specifications if necessary.

Some companies and designers design a board regardless of cost, either because of the complexity of the board or the inability to determine initial estimates. Companies that work with budgets know all too well that estimates must be provided to quote costs and provide estimates of time and expenses.

GETTING STARTED USING TOOLS OF THE TRADE

Board design can't be done with any generic CAD/CAM software. Requirements of the manufacturer dictate a special file format or Gerber files.

Special software is used to design PCBs that help control, design, and check the design for electrical and design requirements. Before choosing a particular brand/version of software, research must be done to find the software, with the best "bang for the buck." Research should include brochures from the company, articles and ratings, testing the software for user friendliness, expansion capabilities, and most important, user forums. User forums provide a method of checking problems within the software, user satisfaction, support quality, and interaction.

Note

Expansion capabilities or program attachments, such as signal integrity and function simulation, are other products that may be used with the standard schematics and PCBs.

UTILITIES AND ACCESSORIES

There are many programs that have bits and pieces of necessary tools that aren't incorporated into layout software. Many of these utilities and programs are in the Designer's Resource (DR ResourceV1.xls) contained in the software bundle that accompanies this book. Programs and utilities such as these reduce manual calculations and provide consistency in the calculations. Some available utilities calculate controlled impedance of a trace, trace resistance in relation to copper thickness and temperature, and so on. Programs available on the Internet require some browsing to acquire but eliminate precious time used in manually calculating these values.

DOCUMENTING STANDARDS AND MATERIALS

One difficult task is to document the available materials for fabricating a board or to receive a listing of materials from manufacturers. Before any choices can be made on what material to use, the available materials must be known.

A designer should document commonly available, or "stocked," materials, by using a table similar to the one shown in Table 5–1. There are several different types of materials, and a table for each type used should be kept. The necessary information to document is the type/name of the material, the Er, and the Tg.

Note

The accompanying Designer's Resource software comes with a material table that records core material with copper thickness and tolerance along with a matrix of which manufacturers carry the material.

GATHERING AND DEFINING PRELIMINARY INFORMATION

This section deals with all the preliminary information used to determine all aspects of the board design. The information is used again throughout the design process to determine other values, materials, and course of action.

Table 5–1 Material Table with Manufacturers

Oz	Core	Oz	Tol.	MFG1	MFG2	MFG3	MFG4	MFG5	MFG6
.5	.006	.5	.001	X	X	X			
1	.006	1	.001	X	X	X			
2	.006	2	.001	X					
.5	.031	.5	.002	X	X	X			
.5	.031	0	.002	X	X				
1	.031	1	.002	X	X				
2	.031	2	.002	X	X				
3	.031	3	.002	X					
4	.031	4	.002	X					

FR-4 Tg=xxx Er=xxx

The designer can't really determine the density of the board or the amount of space required from looking at a simple circuit. The designer must gather all the constraints, requirements, and needs and determine the preliminary set-up (or materials, trace and space, and number of layers).

Designing a PCB Using a Design Checklist

A practical way of explaining the design process is to use a checklist and then define each part of the process as it begins. The following sections incorporate portions of the Designer's Checklist into appropriate sections in a quick overview format, followed by step-by-step explanations. When a checklist doesn't apply, standalone discussions are presented.

Constraints

Constraints include the following:

- Target technology
- Thickness
- Height
- Width

Technology-Driven Constraints

Technology-driven constraints include the following:

- Layer count
- Minimum trace
- Minimum space
- Minimum hole
- Minimum aspect ratio (layer/drill)
- Minimum board thickness

DEFINING CONSTRAINTS AND REQUIREMENTS

Defining constraints is the first key step in designing a PCB. Example design constraints to consider with any design include target technology, thickness, height, and width. Examples of technology-driven constraints include layer count, minimum trace, minimum space, minimum hole, minimum aspect ratio (layer/drill), and minimum board thickness. The initial constraint requirement is an educated estimate of the technology level that should be used for a design. This is only an estimate and may change during the design process if the constraints must be exceeded.

Define Constraints

❑ *Define board dimensions.*

❑ *Define top and bottom board clearance.*

❑ *Note dimensions of cutouts slots and unusable areas.*

❑ *Define the board thickness.*

❑ *Define edge clearance areas.*

❑ *Define all slots and cutouts.*

❑ *Define assembly requirement such as keying information.*

❑ *Mark predefined component locations, including hardware, connectors, lights/LEDs, and switches.*

❑ *Place polygon on mask layer for area that requires no mask.*

❑ *Place keep-out (all layers or per layer) on area that require clearance from/for traces, vias, pads, and hardware clearance (by hardware-to-hole tolerance, hardware movement area, or hole tolerance).*

❑ *Define requirements: IPC, Mil-spec, etc.*

Note

See the following sections, titled "Non-Soldered Thru-Holes" and "Soldered Thru-Holes," for definitions of vias and pads.

Type and Reliability Determination

Selecting the job type and reliability determines, and adds extra constraints, to the entire process. This is the first step of designing a board. If it is determined that Mil-spec, NASA, UL, or a European standard are relevant, investigate further into documentation and requirements into each of those standards before proceeding. This book pertains to a more common industry practice type of design and will not expand on other standards due to the complexity and contradicting information.

If a common industry practice type of board is determined, then the material type is based on the design/assembly requirement only.

Before using any specification in your documentation, you should understand what the specification says and keep a copy handy for discussion. It is possible to design boards without using any specification, by quoting "best commercial practice," but the outcome may be unknown or undesired.

Board Size and Surface Mount Use

Board size and surface mount use are grouped together because of their relationship in the space of the board. Reduction of board size is one of the mainstays of PCB design and is the focus of many board designs. Determining the general space requirements is a preliminary evaluation of the room available.

Designing to an enclosure's requirements and/or designing around the necessary space of a PCB are intertwined. Common personal design philosophies determine which requirements will take precedence. Some have the philosophy that an enclosure should be designed around the PCB's required size, allowing the designer to create "ideal" boards. Others take the position that the board must fit a predefined enclosure or is designed around areas designated for mounting hardware. That determination is the focus of calculating available space. More often, a compromise is made between the required room for components and the necessary size/space of the enclosure and mounting area. Communication between the PCB designer and the mechanical designer is critical at this point.

If there is inadequate room using thru-hole components on the board, then the designer has the following options:

- Place components on both sides.
- Change to surface mount components.
- Use surface mount components on both sides.
- Increase board area.

There is little documentation, calculations, or programs that can surpass experience in determining available room and proper location. The options are as follows:

- Thru-hole components versus SMT (surface mount technology) components
- SS (single-sided) board versus DS (double-sided) board
- DS versus ML (multi-layer)

This common compromise among component size, board size, and enclosure size is exactly what forced the progression of technology into creating surface mount technology and multi-layer boards.

Noting RF/EMF Considerations

RF (radio frequency) and EMF (electromagnetic field) are the two major forms of external interference. At this time RF and EMF considerations are only noted, since there is little that can be done at this point.

RF in layperson's terms is very high frequency causing/injecting interference in a signal line or trace. EMF is an electromagnetic field caused by a combination of frequency and a coil producing magnetic type interference. These considerations are usually from external sources and sometimes may be from components mounted to the board. At this point only external RF/EMF considerations are noted.

Environmental Considerations

Heat, vibration, humidity, radiation, and other factors need to be taken into consideration in determining the components, coatings, material, type of reliability the board will have, and if the board can be manufactured.

Very rarely will the heat in the environment will affect the board, but the components are more susceptible to environmental conditions.

Vibration is more a determination of effects over time as opposed to changing any assembly process.

Humidity causes corrosion and possible shorts (in the right conditions). The board can be protected from many environmental conditions by coatings, such as solder masks and conformal coating.

Defining the Required Board Area

The area for the physical board is a concern, as is the room for the components. This creates a three-dimensional aspect in the design.

The board should not be exactly the same size as the available room. There is a tolerance, or a ± value, when the board is manufactured, and this should be accounted for. If there is no required board size, the board size will be determined by balancing aesthetic/economic board size, required surface area for parts mounted to the board, and design area to accommodate the number of traces balanced against the number of layers.

Defining the Required Board Thickness

Either a board thickness is defined by its enclosure/mounting hardware or the mounting hardware is defined by the required board thickness. The best situation is a compromise between the two.

Component thickness or component selection becomes a compromise. Component variations are primarily selected starting with required board thickness. A component with substantially long leads isn't much of a problem, but leads that do not extend at least the width of the hole diameter past the board are a concern and a problem. Therefore, if a component with adequate lead length is unavailable, the board thickness must be reduced, or an alternate component must be used. If a board thickness isn't required by a enclosure or mounting equipment, then the board thickness is determined by selecting from standard thickness (.031″, .062″, .125″, etc.) and the shortest lead length available for any of the components used.

Tip

The board thicknesses noted are U.S. standards; thickness standards may be different in other countries.

❏ *Determine assembly type for production, such as manual, automatic, or manual prototype-to-automatic.*
❏ *Determine servicing type:*
  ❏ *No service/troubleshooting (throw-away board)*
  ❏ *Low service (inexpensive components on the board, or easily swappable application, low serviceable location)*
  ❏ *Highly serviceable (expensive components on the board or difficult to swap application, highly serviceable location)*
  ❏ *Determine technology limitations and target technology*

DETERMINING THE MATERIAL TYPE TO USE

There are several different types of material available, along with other materials that are only variations on the four or five basic material types. There are several questions to ask to determine the type. Heat, impedance, and flexibility are the basic determinations of what material to use. Ask these questions for every job and/or application:

• Is there a large amount of components (especially thru-hole) to be mounted on this board? Soldering components to or on the board is a major factor in the ratings.
• Are there any flexibility requirements for this board? Flexibility requirements limit the amount of material types available to use, since most are in the "rigid" material category.

- Will the board be auto assembled, or is there a chance of it being auto as-sembled in its lifetime? Prototypes often will not be auto assembled. So if there is a chance for auto assembly, then heat is a large factor.

See Chapter 3 for information on selecting material based on assembly re-quirements.

DESIGNING THE BOARD

With the initial requirements and constraints determined, several choices must be made. A compromise is made between the available materials, the required cur-rent requirements, board thickness, and an estimate of the number of layers and trace and space.

- ❏ *Open new file or load appropriate template.*
- ❏ *Check for standards in pads, vias, or text styles.*
- ❏ *Draw board border using .040″ line on center.*
- ❏ *Draw all slots/cutouts in board using .040″ line on center.*
- ❏ *Enter design information.*
- ❏ *Load libraries/archived library.*
- ❏ *Load netlist.*
- ❏ *Generate BOM and compare against parts list. (This is to include mechani-cal components not in the schematic.)*
- ❏ *Place parts with a predefined location where necessary.*
- ❏ *Define classes/nets with trace width, clearance (space), and hole clearance.*

Selecting Material Thickness and Copper Weight

Determining which material to use for double-sided boards is simple enough. Standard materials come in two basic configurations: multilayer core and double-sided core. ML core material is measured by core material separate from the cop-per thickness. DS core material is measured overall, including the copper thickness. Figure 5–1 displays a board stack-up that is 10% less than the available space in the enclosure shown in Figure 5–2.

Note

The term *stack-up* refers to the representation of the order the materials placed. *Lay-up* refers to the physical placement of the material preformed by the manufacturer.

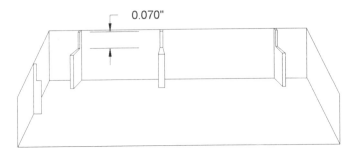

0.070"

Figure 5–1   Board thickness 10% less than max.

Determine Copper Thickness

The copper thickness can be selected by initial copper thickness and design to the constraint (thickness) or by determining the thickness requirements from the circuit current/voltage requirements and some of the controlling space requirements. Figure 5–3 displays spacing constraints, one of the controlling factors/limitations for a sample board.

After some general routing or knowledge of the board is known, a determination of some of the trace requirements can be made. If possible, use 1/2 oz copper to start with and go from there.

An example circuit is only 5 V but carries 3 Amps of current. 5 V only requires .001″[.0254] of space, but the technology constraints require a minimum of .006″[.1524] (both sides). The spacing requirements for Figure 5–3 are only .020″[.508]. Subtracting .012″[.3084] (.006″[.1524] on either side) only leaves .008″[.2032] for a trace. As shown in Table 5–2, a .008″[.2032] trace at 1/2 oz can only handle a little over .2 Amps, so the copper thickness must be increased to 1 oz.

Another option would be to reduce the pad size, if the servicing requirements allowed. In this example, the pads are at their minimum allowance.

Spacing between the leads and pads of a component is one of the common determining factors in copper thickness and technology.

TABLE 2: MATERIAL STACKUP

LAYER	MATERIAL	THICK	TYPE
TS			
TM	MASK	.0025″	
L1	COPPER	.0021″	TRACE
	DIELECTRIC	.0100″	
L2	COPPER	.0007″	PLANE
	DIELECTRIC	.0140″	
L3	COPPER	.0007″	PLANE
	DIELECTRIC	.0100″	
L4	COPPER	.0021″	TRACE
BM	MASK	.0025″	
	TOTAL	.062″ (+/−.010″)	

Figure 5–2   Dimensions of room for board thickness.

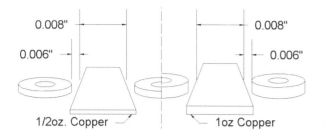

Figure 5–3   Spacing between pads.

Note

The accompanying software allows the designer to calculate the pad size and, using pad spacing, trace width, and spacing, show the number of traces allowed between the pads.

Defining Trace/Width

Whether a schematic was completed or not, the circuits or nets must be defined for current and voltage. Current is relative to a square inch of copper for a particular temperature. The square inch of a trace is defined by a trace's width and the thickness of the copper it is etched from. The third determining factor is the temperature increases above room temperature. First a desired temperature is selected. This is a temperature that is desired for an "across-the-board" value. If it causes no increase in layers or square inch required for the board, 0°C should be selected. A compromise between available copper thickness and available trace width is made based on combinations available for the temperature requirements.

Table 5–2   Copper/trace requirement per ounce copper

Trace	1.5 oz Cu External Amps	.5 oz Cu Internal Amps	1 oz Cu Internal Amps
0.001	0.23	0.05	0.08
0.002	0.37	0.08	0.14
0.004	0.62	0.14	0.23
0.005	0.73	0.16	0.27
0.006	0.83	0.19	0.31
0.008	1.02	0.23	0.38
0.010	1.20	0.27	0.45

For all jobs, the initial trace selection should be a considerable amount more than what is necessary and above that of the minimum amount for the technology. An example would be for the advanced technology, which is a .006"; therefore, a .008" trace would be preferred, allowing for reduction when necessary. It is not necessary to use the minimum trace width for the current rating unless space is tight.

Current and square inch are not linear due to the increase of heat. This means that there is no simple multiplier for the calculation.

Standardizing Trace Width

Standardizing trace width may seem bizarre, but ease of route and consistency are a large payoff in the end. When spacing is a concern, the designer will know the necessary room to route several traces because of familiarity with trace width, clearance, and multiple trace width. Although external and internal traces and clearances can be different, unless space is very tight, a worst-case scenario may be used. As mentioned earlier, less space is necessary internally due to higher insulation, but more trace width is necessary to dissipate heat since the internal trace is enclosed; thus a compromise may be reached. For the worst-case type routing:

- Use the internal trace width.
- Use the external clearance.

(This practice is fairly common, and "define by layer" attributes are not available in some lesser quality software.)

Grid routing is also extremely important. Many designers continue to prefer a more aesthetic board, which may require grid routing. For inch grids, multiples of 6 are quite common. *Example:* .025" grid using a .012" trace leaves a .013" clearance, or effectively 1:1 ratio (12/12).

If using a .001" grid, .006" is the common minimum trace, for 1/2 oz copper. (Refer to Table 5–4 for minimum t/s per technology/copper thickness.)

Selecting the Dielectric Material

Copper thickness is initially selected, and then dielectric material is selected based on availability. These are the values that determine core thickness:

- Overall thickness
- Pre-Preg material thickness
- Available materials
- Copper thickness

Simultaneously, the copper thickness, Pre-Preg, and core thickness per layer can be calculated to determine what cores available may be used as shown in Figure 5–4.

Inches		Difference		0.0954
		Target		0.1250
		Total		0.0296
Use for all Dielectrics-->				0.0080
Use for all Coppers-->				0.0007

Layer	Grafix	Type		Thick
	-----------	Plating	▼	0.0014
1	-----------	Signal	▼	0.0007
	~~	Pre-Preg	▼	0.0080
2	=======	Plane	▼	0.0007
	~~	Pre-Preg	▼	0.0080
3	-----------	Signal	▼	0.0007
	XXXXXX	Core	▼	0.0080
4	=======	Plane	▼	0.0007
		Plating	▼	0.0014

Figure 5–4   Calculating core, copper, dielectric, and overall thickness.

Note

In the accompanying software, stocked material may be recorded to determine material use.

Figure 5–5 provides an example material table used to track available thickness, configuration, and tolerance per material type.

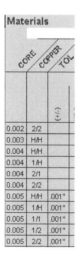

Materials		
CORE	COPPER	TOL
		(+/-)
0.002	2/2	
0.003	H/H	
0.004	H/H	
0.004	1/H	
0.004	2/1	
0.004	2/2	
0.005	H/H	.001"
0.005	1/H	.001"
0.005	1/1	.001"
0.005	1/2	.001"
0.005	2/2	.001"

Figure 5–5   Core material and copper thickness.

Defining Copper Thickness, Trace Width,
Number of Layers, and Technology

All designs hinge around the current and voltage requirements and the components used. The minimum trace and clearance is not desired because it will usually change/increase technology and cost. The attempt is to stay with a lower technology while keeping the design routable. The board may be populated with components before defining the width and clearance. How small the width and clearance needs to be depends on the clearance between pins on many of the components. Most of the decisions will be made from experience, but without experience, trial and error is next.

Anytime during the design, the number of layers may be increased or decreased. If any of these elements falls below the limits of the technology and falls into a higher technology, the entire design may as well be within that technology.

With that in mind, the width and clearance can be defined and routing attempted. If the routing fails, then either the layers or the technology should be increased. By increasing the number of layers (a much simpler choice), no other attributes need be increased and the design continues. If increasing the number of layers is not an option or the design is already at the maximum number of layers for that technology, then the technology should be increased.

By increasing the technology, the following options may change:

- Width decreases
- Clearance decreases
- Pad diameter decreases
- Aspect ratio increases
- Via holes size decrease
- Via pad size decreases
- Number of layers increases

It is obvious that either initially determining the correct technology or increasing the number of layers is very important and time saving.

Determining material and copper thickness and trace widths becomes less complicated after time when standards are set and common stack-ups are determined. Most boards a designer will make will use 1/2 oz copper material for all layers. This is the most common material and has the greatest number of different thicknesses. Keeping this in mind, most designs will start with these values. Board thicknesses of .031 and .062 are the most common because the "core" materials stocked are in .031 multiples (.031, .062, .093 etc.). If not a single-sided or double-sided board, multilayer boards are either built to a desired thickness or to one of the .031 multiples. Designing to these values provides consistency and standardization. Using these values, all trace width and clearance calculations can then be determined. Just because a trace width can be .006″ or .005″ doesn't mean it should be designed that way.

Trace width and clearance should initially be at least .002″ larger than the smallest width and clearance for that technology. This will allow the designer to decrease the width and clearance to the technology limit before changing technologies.

Selecting the component with the densest pattern, such as a PGA, BGA, or a dense connector, and then attempting to route this area first can help determine the width and clearance before too much work is done. Often components such as these will determine the technology of the board. Table 5–3 is an expandable quick list of components and the technology of the board used. These are generalities and, like most generalities, have exceptions. For SMT, the generality is that if the designer requires SMT, then room is most likely a concern. Although SMTs may be used on conventional technology boards, advanced is usually used. Components such as BGA(s) usually require leading edge because of the number of pins that are grouped together and the need for the traces to exit the area.

For example, a BGA that has six rows to the center in advanced technology may require at least five signal layers to exit the component area, whereas a leading-edge technology board would allow half that amount.

Many components require a certain level of technology if they are used. Table 5–3 shows some common components and the required technology use.

Table 5–4 shows a quick list of the technologies. Table 5–5 shows a table of changes to be made when increasing the technologies.

❑ *Define other design attributes or design rules.*
❑ *If applicable, define class trace and space by layer.*
❑ *Configure design/job for*
  ❑ *Overall design rules*
  ❑ *Mask swell (global)*

Table 5–3   Quick List of Components and Technologies

Components	Min. Technology
Thru-Hole	Conventional (.006″/.006″)
SMT	Advanced (.005″/.005″)
BGA, PGA	Leading edge
	State of the art

Table 5–4 Quick Reference Guide Limits per Technology

Attribute	Conventional		Advance		Leading Edge	
	Use	Minimum	Use	Minimum	Use	Minimum
Min. Finished Drill	.014"[.3556]	.012"[.3048]	.012"[.3048]	.010"[.254]	.010"[.254]	.008"[.2332]
Min plated hole for (based on .005" plating)						
.031" board	.013" [.3302]		.007"[.1778]		.003"[.0762]	
.042" board	.013" [.3302]		.007"[.1778]		.003"[.0762]	
.080" board	.013" [.3302]		.007"[.1778]		.005"[.0762]	
.100" board	.013" [.3302]		.008"[.2332]		.008"[.2032]	
Clearances (starting copper)						
Clearance (.5 oz)	.008"[.2032]	.006"[.1524]	.006"[.1524]	.005"[.127]	.004"[.1016]	.003"[.0762]
Clearance (1 oz)	.010"[.254]	.007"[.1778]	.006"[.1524]	.006"[.1524]	.006"[.1524]	.004"[.1016]
Clearance (2 oz)	.012"[.3084]	.008"[.2032]	.010"[.254]	.007"[.1778]	.008"[.2032]	.005"[.0762]
Clearance (3 oz)	.014"[.3556]	.010"[.254]	.012"[.3084]	.008"[.2032]	.010"[.254]	.008"[.2032]
Plane to edge	.020"[.508]	.008"[.2032]	.010"[.254]	.006"[.1524]	.008"[.2032]	.003"[.0762]
Hole clearance (SS/DS) +electrical clearance	.008"[.2032]	.006"[.1524]	.006"[.1524]	.004"[.1016]	.004"[.1016]	.002"[.0508]
Hole clearance (ML)						
+electrical clearance	.010"	.009"	.008"[.2032]	.006"[.1524]	.005"	.003"
Lead clearance	.010"		.010"		.010"	
Width (.5 oz)	.008"[.2032]	.006"[.1524]	.006"[.1524]	.005"	.004"	.003"
Width (1 oz)	.010"	.007"	.008"	.006"[.1524]	.006"[.1524]	.004"
Width (2 oz)	.012"	.008"[.2032]	.010"	.008"[.2032]	.008"[.2032]	.005"
Width (3 oz)	.014"	.010"	.012"	.010"	.010"	.008"[.2032]
MFG AR (SS/DS)		.006"[.1524]		.004"		.002"
MFG AR (ML)		.009"		.006"[.1524]		.003"
Pad dia. via (SS/DS)	Hole + .024"	Hole + .022"	Hole + .020"	Hole + .018"	Hole + .016"	Hole + .014"
Pad dia. via (ML)	Hole + .030"	Hole + .028"	Hole + .024"	Hole + .022"	Hole + .018"	Hole + .016"

Attribute	Conventional		Advance		Leading Edge	
	Use	Minimum	Use	Minimum	Use	Minimum
Pad dia. soldered–PLTH (Ideal)	$2 \times$ hole		$2 \times$ hole		$2 \times$ hole	
Pad dia. Soldered–PLTH (Mid)	$1.75 \times$ hole		$1.75 \times$ hole		$1.75 \times$ hole	
Pad dia. Soldered–PLTH (Min.)	$1.5 \times$ hole		$1.5 \times$ hole		$1.5 \times$ hole	

Note: These values are based on the technology table in the DFM section. Adjust them to accommodate individual findings. The "Use" column is suggested values to use. It is recommended not to use the minimums unless necessary.

❑ *Paste swell (global)*
❑ *Plane swell (global)*
❑ *Pad swell (global)*
❑ *Thermal divide (pad dia./4)*
❑ *Thermal clearance*

The Pad and the Thru-Hole

Understanding the manufacturing and assembly constraints and considerations allows the designer to determine the proper pad surface area. IPC has specifications for minimum annular ring, but these are minimum possible values and should not be used unless necessary. IPC specifies .002″ external and .001″ internal minimum annular ring. This value should be over the hole wall. The AR shouldn't be the same size as the hole wall plating, since the external pad behaves like a cap for the hole wall and helps ensure that the hole wall will stay in place. With most

Table 5–5   Technology Quick-Change Table

Technology change	Width reduction	Clearance reduction	Pad reduction (MFG AR)	Aspect ratio reduction	Via pad reduction	Layer increase
Conventional to Advanced	.001″ (.006″ to .005″)	.001″ (.006″ to .005″)	−.003″	8 : 1 to 10 : 1	−.003″ or new via	10/12 to 20 layers
Advanced to leading edge	.002″ (.005″ to .003″)	.002″ (.005″ to .003″)	−.002″	10 : 1 to 12 : 1	−.002″ or new via	20 layers to 30 layers

Note: Multiply by 25.4 for metric values.

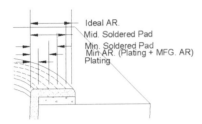

Figure 5–6   Min. AR regardless of specs. Pad divided into its parts (with manufacturer minimum AR).

technologies the manufacturer requires an additional pad so the hole wall is guaranteed to have some pad area, as shown in Figure 5–6.

Defining the Thru-Hole

Holes and thru-holes have been traditionally broken up in two groups: plated (supported) and non-plated (unsupported) holes. The term *supported* refers to the plating in the hole wall. Non-plated or unsupported holes may or may not have a pad such as a mounting hole and no hole wall plating (Figure 5–7). This is manufacturing terminology but for designing the holes should be broken up in two categories of soldered and non-soldered.

In each of those categories the classification of plated and non-plated should be identified.

- Soldered
  - Plated thru-hole (PLTH) (including vias)
  - Non-plated thru-hole (NPTH)
- Non-soldered
  - Plated thru-hole (PLTH)
- Non-plated thru-hole (NPTH) with and without pad

The designer must know if the pad is identified as soldered or non-soldered first. This information helps the engineer to determine if the pad calculation should be for soldering or for the minimum annular ring. If the pad is unsoldered, a standard AR can be applied and doesn't change according to assembly requirements.

Note

A via is simply a plated thru-hole that isn't soldered. It does not require a calculation to determine pad size per assembly or application, but rather a simple calculation of the minimum annular ring or the smallest, most cost-effective size that can carry adequate current.

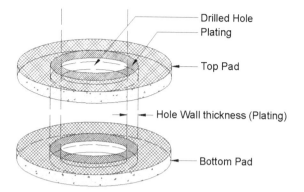

Drilled Hole
Plating
Top Pad
Hole Wall thickness (Plating)
Bottom Pad

Figure 5–7   Thru-hole definitions.

Non-Soldered Thru-Hole.   Minimum annular ring (or Min. AR) comprises two different terms. A manufacturer, when talking about a Min. AR, is speaking of the Min. AR required by the standards or specification the designer has noted, plus its own AR requirements. The specified annular ring, from IPC or other standard, is the minimum AR required for board completion. Table 5–6 displays the minimum possible AR regardless of specifications. This value should not be used but rather a more realistic value.

This value is for vias or non-soldered pads and should not be used with soldered pads. Remember that the AR is only 1/2 of the pad diameter. To accomplish this, the manufacturer requires additional pad area to account for the errors of its process: imaging, drill, and registration, as shown in Figure 5–6 and Table 5–7. In conventional technology each of these is about .003″ Therefore, the manufacturer's AR must be added to the pad:

Non-soldered total AR = finished AR + manufacturer's AR
Example: .0025″ + .009″ = .0115″ Total AR

To find the total diameter, the total AR of this value must be multiplied by 2:

Total diameter = total AR × 2
Example: .0115″ × 2 = .023″

Now that the pad requirements have been found, the finished hole diameter must be added to find the overall pad diameter:

Finished pad dia. = total diameter + finished hole diameter
Example: .023″ + .008″ = .031″

Therefore, the *minimum* pad diameter for a .008″ hole should be a .031″ pad. Again, it must be noted that this is a minimum and shouldn't be used unless necessary, or the manufacturer used is very consistent. The recommended size

Table 5–6   Finished Annular Ring (without MFG AR)

	Plating	+	Spec. AR	=	Min. AR
External pad	.0025″	+	.002″	=	.0045 round to .005″
Internal pad	.0025″	+	.001″	=	.0035 round to .004″

would be at least .002″ over the minimum diameter. These example values were based on conventional technology.

Find internal/external pad dia = Min. AR (.0025″) + MFG AR (.009) × 2 + hole dia. Example: (.0025 +.009″ × 2 + .070″)

The AR should be, at the minimum .002″ (annular) larger than the plating for the external pad and, at the minimum .001″ larger that the plating for an internal pad. Effectively the plating thickness is .0025–.0030″ (specify which to use in your fabrication notes). Use Table 5–6 as a rule of thumb for the AR before adding the MFG AR.

Now the manufacturer annular ring is added (Table 5–8) and becomes part of the pad. The MFG AR is added to account for the manufacturer's errors. These must be added to display the finished pad size.

Using these values, the designer may simply add the hole, and the result is the finished pad.

Table 5–9 is an example using a .031″ hole.

Soldered Thru-Holes.   Most of the same rules apply to the soldered thru-hole except that the external surfaces must be larger to dissipate heat to avoid problems described in Chapter 3. If possible, for consistency and ease of calculation, the internal pad should be the same as the external, unless a reduced pad is necessary. The reason for this is that the external pad is soldered and the internal pad is not. A soldered thru-hole must also increase in proportion to the lead diameter since a larger lead requires more heat; thus to distribute heat evenly between the lead and the pad, the pad must increase as well.

Table 5–7   Generic pad calculation

AR	+	Mfg AR	× 2	+ Hole	= Pad Dia.
.0025″	+	.009″	× 2		
.0025″	+	.009″	× 2		

Note: These values are based on conventional technology.

Table 5–8   Generic Finished AR (including Min. AR and MFG AR)

	Min. AR	+	MFG. AR (Conventional/ advanced/ leading edge)			×	2	=	Pad Area (conventional/ advanced/ leading edge)		
			Conv.	Adv.	L.E.				Conv.	Adv.	L.E.
External (DS)	.005	+	.006″	.003″	.002″	×	2	=	.017″	.011″	.009″
External (ML)	.005	+	.009″	.006″	.003″	×	2	=	.023″	.017″	.011″
Internal (DS)	.004	+	.006″	.003″	.002″	×	2	=	.016″	.010″	.008″
Internal (ML)	.004	+	.009″	.006″	.003″	×	2	=	.022″	.016″	.010″

The pad for the soldered thru-hole has a range of acceptable size. That range is no less than the minimum AR up to 2 times (2 ×) the hole size. The rule of thumb for a soldered pad is 2 times the finished hole diameter. Three basic multipliers, which correlate with design requirements, can be used to determine the pad diameter (Table 5–10). Figure 5–8 can be used to determine which calculation to use.

Note

These values are based on the standard component lead material and may change according to material type.

Note

The Designer's Resource software calculates these three values and will display values no less than the minimum AR (includes finished AR & manufacturer AR).

Note

The accompanying software allows you to adjust the "over hole value" (+.010″ by default) as well as adjust pads, across the board, for special con-

Table 5–9   Quick Pad Dia. Table (conventional/advanced/leading edge)

	Pad Area	+	Hole Dia.	=	Finished Pad
External (DS)	.017/.011/.009″	+	.031″	=	.048/.042/.040″
External (ML)	.023/.017/.011″	+	.031″	=	.054/.048/.042″
Internal (DS)	.016/.010/.008″	+	.031″	=	.047/.041/.039″
Internal (ML)	.022/.016/.010″	+	.031″	=	.053/.047/.041″

Table 5–10   Calculating the Finished Soldered Pad Diameter

Finished Hole Size	× 1.5	= Minimum finished soldered pad diameter	(no less than finished hole + AR +MFG AR)
Finished Hole Size	× 1.75	= Median finished soldered pad diameter	(no less than finished hole + AR +MFG AR)
Finished Hole Size	× 2	= Maximum finished soldered pad diameter	(no less than finished hole + AR +MFG AR)

siderations, and adjust the pad by the copper thickness selection. The printable forms include worksheets for calculating hole and pad.

The Thermal Pad.   Thermal connections are features used/required for only soldered pads. Normally used on plane layers or in a solid copper area, they are three or four traces connected to the pad in a + or × shape. The actual traces are known as spokes because of their bicycle spoke appearance. Instead of connecting the plane/solid copper area directly to the hole wall, these spokes are used to reduce the heat sinking affect of the copper areas. A balanced ratio of board copper to component lead metal dictates that the soldered thru-hole have limited

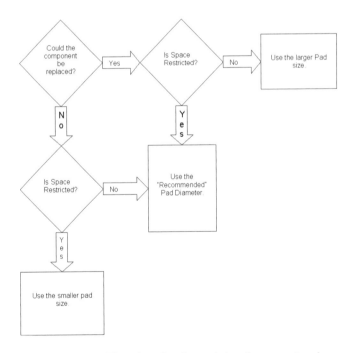

Figure 5–8   Flowchart for determining the correct pad.

copper area, but maintaining adequate area for current-carrying capacity. For this reason the combination of the spoke widths is equivalent to the pad diameter. The thermal pad is the same diameter as other inner-layer pads. The area containing the pad and the spokes is known as the thermal outer diameter. This distance is the inner diameter × 1.5.

The term *thermal pad* is now a misnomer. The hand tape method used negative planes, and an actual pad was placed. Now with computer-generated images, the pad is no longer used.

Non-Plated Thru-Holes.   There are two types of non-plated thru-hole; with a pad and without a pad. It is important to note that a pad is either plated or nonplated. During the plating process, if a pad exists, the hole will plate with copper. If a pad should not be plated, then it must be drilled after the plating process, creating an extra process and additional cost. The NPTH without a pad may be drilled at the same time as the others.

Non-Plated Thru-Holes, with Pad.   Non-plated thru-holes (NPTHs) have no plating in the hole that is drilled through the pad. That means there is no additional support to hold the pad to the board besides the normal copper adhesive. For this reason the pad must be larger to help adhere or hold the pad to the board if heated or soldered. IPC specifies that such unsupported pads should have an annular ring of .006″[.152]. This value should be even larger if the pad is soldered.

Non-Plated Thru-Holes, without Pad.   The generic NPTH is no more than a hole in a board, with no pad or hole wall plating. These are used for numerous reasons, such as a mounting hole (or screw holes), accessing holes for screw adjustments, or a routing hole for wire. No plating or annular ring requirements are necessary. The general thru-hole is unlike the other holes because it has no plating or soldering considerations.

Mounting Holes

Clearances around mounting holes are larger to compensate for drill tolerance. These holes are not the same as board-to-edge to regular clearance. It is effectively clearance requirements + drill registration + layer registration (if applicable). The drill registration error and the layer registration error are still there, but there is not an image registration, since there is no image. Mounting holes need to be treated differently because of the lack of pads. The manufacturer's annular ring compensates for the misregistration; thus all the errors are retained in the pad area. This means the annular ring, usually reserved for pads, is converted into clearance area.

If the software allows keep-outs, clearance areas around the pads, or some sort of mounting hole clearance attribute, it should be the same as the manufacturer's ring as a general rule.

Aspect Ratio

As explained in Chapter 2, first the minimum hole available must be determined, otherwise a manufacturer will not be able to build the board or the board may be very expensive to build. As mentioned earlier, when discussing capabilities with the manufacturer, use the terms *starting* or *finished* aspect ratios and *minimum finished drill.* Table 5–11 provides a quick reference to finished aspect ratios.

Tip

If the manufacturer can only provide the starting aspect ratio, then find the minimum starting hole and add the "drill over," or the amount the manufacturer drills over the finished hole. Then divide this into the same board thickness to provide the finished aspect ratio (for example, 12:1 starting aspect

Table 5–11   Quick Table of Finished Aspect Ratios

Board Thickness	For 5:1 (conventional) Min. drill .018″	For 8:1 (advanced) Min. drill .012″	For 10:1 (leading edge) Min. drill .008″
.010	.002	.001	.001
.020	.004	.002	.002
.030	.006	.003	.003
.040	.008	.005	.004
.050	.010	.006	.005
.060	.012	.007	.006
.070	.014	.008	.007
.080	.016	.010	.008
.090	.018	.011	.009
.100	.020	.013	.010
.125	.025	.016	.012
.150	.030	.019	.015
.175	.035	.022	.017

Note: Shaded areas are below minimum drill size. These values are based on the average minimum drill for manufacturers (per technology). Values may vary by manufacturer, but using the average size ensures the ability for all manufacturers of a similar technology to fabricate the board.

ratio, in a .125″ board = .010″ drill). Add drill over [.005″] = .015″; now divide into the board thickness = 8. This is the first part of the aspect ratio 8:1.

Determining What Fabrication/Registration Errors Are Applicable

Although it is best to use the AR value as a minimum with all pads, it becomes an issue when space is a problem. The annular ring consists of

- Drill registration tolerance
- Image registration tolerance
- Layer registration (For multilayers)

These values control many aspects of the board, including

- Board edge
- NPTH pad diameter
- PLTH pad diameter
- Trace to hole clearance
- Trace to cutout clearance

The following images display the affects of image, layer, and drill shifts during the process, accounting for the manufacturing annular ring. Figure 5–9 displays a normal board with the image (copper pad), layer, and drill aligned.

Figure 5–10 is an exploded view of the board displaying each feature of the three features.

Figure 5–11 displays the effects of a layer shift. In this example the inner layer is shifted within an acceptable amount.

Figure 5–12 displays the image shifted in addition to the same layer shift.

Figure 5–13 displays a shift in the drill in addition to the previous image and layer shift.

Each shift is within acceptable standards (+.003″). Each shift accounts for .003″ and in combination add up to .009″ of shift. These allowable shifts define the manufacturer's annular ring.

Image registration tolerance is almost always there to some extent. Direct imaging greatly reduces the value, but some will remain.

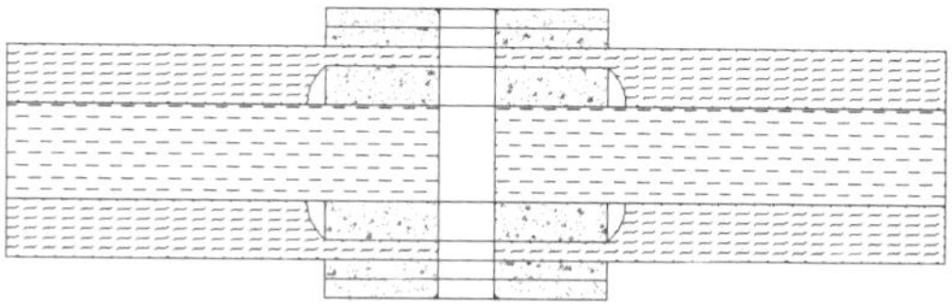

Figure 5–9   Multilayer board.

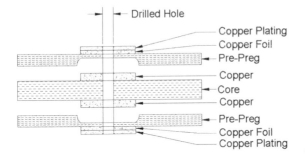

Figure 5–10   Multilayer board exploded.

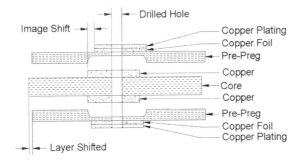

Figure 5–11   Multilayer board with a layer shifted during press process.

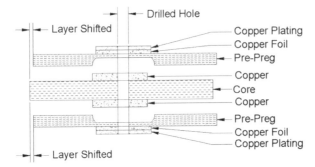

Figure 5–12   Multilayer board with a layer and image shifted.

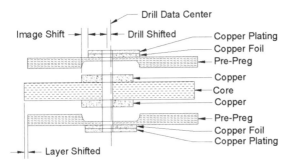

Figure 5–13   Multilayer board with layer, image, and drill shifted.

Lay-up registration is calculated only in multilayer boards. This registration is incurred when multiple layers are stacked together and pressed. For a single/double-sided board, the board is one single piece and cannot move during the process, thus eliminating any registration error.

AR specifications usually includes the layer registration, so when a SS or DS board is done, simply remove that value. This is why it is important to have your manufacturer define these values.

Finding the Current Capacity of a PLTH

To determine the current-carrying capacity of a PLTH, the hole must be laid flat to provide the equivalent of a trace (see Figure 5–14 for an example). This will not be exactly equivalent, but by using the finished hole the value is more than adequate.

$$\text{finished hole} \times \text{Pi} = \text{Width}$$
$$\text{Example: } .030 \times 3.14 = .0973$$

The plating thickness is then matched to an equivalent copper thickness. Common hole wall thickness is .0025–.0028″, equivalent to a 2 oz copper. Calculate the current capacity of a 2 oz copper × .0973″ wide.

$$.0007'' = .5 \text{ oz, } .0014'' = 1 \text{ oz, } .0028'' = 2 \text{ oz, } .0042'' = 3 \text{ oz}$$

Defining Space/Clearance

Space requirements are defined by either voltage requirements or for mechanical error compensation, or both.

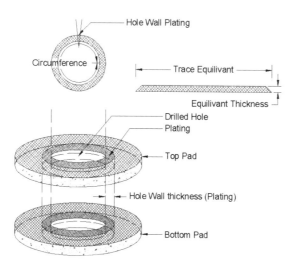

Figure 5–14 Hole wall plating and trace equivalent.

In regard to electrical requirements there are clear multipliers for the internal and external layers. This is a simple calculation and a simple process. In regard to mechanical error compensation, clearances are in addition to the electrical clearances.

Trace to Trace and Trace to Via. These values are approximate to that of the technology requirements. This clearance should be the minimums of the technology requirements or the electrical requirements. Since neither the trace or the via is soldered, there are no solder clearance considerations. Values such as Trace-to-(soldered) Pad and Trace-to-Hole depend on additional requirements, such as registration and soldering clearance.

Trace to Pad (Soldered). Trace to pad in general requires only the same clearance as the electrical clearance. Because of solder mask registration errors, or "solder mask swell," the external clearances between the soldered pad and the trace need to be larger. This eliminates the possibility of a trace being exposed and possibly shorted when the pad is soldered. In addition, the electrical clearance requirements are not from trace to trace and trace to pad, but also to exposed areas. This requires the electrical clearance to be added to the trace to pad clearance.

The typical setting for a conventional board is .010"[.254] over on mask swell (.005"[.127] per side) and the minimum clearance is .006"[.1524] Average registration error for solder mask is .003"[.0762]. Adding the .003"[.0762] with the .005"[.127] of the mask swell allows a .008"[.2032] of area exposed around a pad.

This then dictates that the Trace-to-Pad clearance is the annular mask swell + mask registration + electrical clearance (as shown in Table 5–12).

Trace-to-Hole. Trace-to-Hole clearance is similar to Trace-to-Pad clearance since it differs from just the standard clearance. A pad and a trace are on the same layer, and when the layer or film is misregistered, both the pad and the trace move together. With a pad and a hole, the manufacturer requires additional room in the pad to account for misregistration between the hole and the layer(s). Thus any registration error is retained within the pad and, internally, the clearance is only the electrical clearance. With the trace to hole clearance, since the hole isn't retained in the pad, the clearance includes the registration error plus the electrical clearance and the plating, since the actual drilled hole is where the measurement is taken from. Table 5–13 may be used to calculate Trace-to-Hole distance.

Table 5–12   Trace-to-Pad clearance

Annular mask swell	+	Mask Registration	+	Electrical Clearance	=	Clearance

Table 5–13   Trace to Hole (SS/DS)

Plating	+	Registration error (layer error + drill error + image error)	+	Electrical Clearance	=	Clearance

Only those registration errors that apply need be added. An SS/DS board doesn't have a layer registration error. Only ML boards are pressed together have a possible lay-up error.

An example of a ML board, using conventional technology with a .006″ electrical clearance, would look like that shown in Table 5–14.

An example of a DS board, using conventional technology with a .006″ electrical clearance, would look like that shown in Table 5–15.

Hole-to-Hole.   Hole-to-hole clearance is simple. Same size holes are drilled at the same time and would only need to be separated by the spacing shown in Table 5–16. Because of the tolerances in drilling, drilled holes have the chance of "breaking out" or breaking into each other's space. These values depend on the technology of the drill used, plus additional space to provide additional strength.

Pad to Pad.   Pad-to-Pad is not only limited by technology but by the assembly requirements. Large pads that consume large amount of solder and heat usually require a larger soldering iron. This forces clearances on soldered areas to be more as shown in Table 5–17.

Solder Dams

A solder dam is material placed between pads to prevent solder flowing from one pad to another. This is a consideration usually with surface mount components with a fine pitch, or when the pads are very close together. Solder mask or some other type of coating separates large pitch surface mount components. The small

Table 5–14   Trace to Hole (ML board)

Plating	+	Registration error (lay-up error + drill error + image error for ML)	+	Electrical Clearance	=	Clearance
.0025″	+	.003″ + .003″ + .003″	+	.006″	=	.0175

Table 5–15   Trace to Hole (SS/DS)

Plating	+	Registration error (drill error + image error for DS)	+	Electrical Clearance	=	Clearance
.0025"	+	.003" + .003"	+	.006"	=	.0145

gap between surface mount pads and mask swell allows very small amounts of mask to be placed between pads. These limits are controlled by the technology of the manufacturer and the manufacturer's limits. Solder mask minimum width is usually the same as the minimum trace and space. When using a very high technology, such as leading edge or state of the art, different materials may be used for masking when chip size and component size become excessively small.

❑ *Define assembly direction (especially auto assembly).*
❑ *Define/determine component direction. For high service boards all I.C. should be in the same direction and oriented the same way.*
❑ *Define areas by type.*
❑ *Define layers, including.*
   ❑ *Number*
   ❑ *Symmetry (signal plane signal, etc.)*
   ❑ *Layer direction*
   ❑ *Layer type (strictly power, digital, etc.)*
   ❑ *Split planes*
❑ *Copy board, cutout, and slot outlines to all plane layers providing copper to edge clearances.*

Clearances and Board-to-Edge Clearance

Copper-to-Edge clearance, also known as Board-to-Edge clearance, is a line twice the width of the required clearance that may be used for the border and clearance areas on plane layers. The fabrications notes should instruct the fabrica-

Table 5–16   Hole-to-Hole Clearance

Drill Size	Minimum Spacing Requirements (Conventional)
<.030"	.010"
<.080"	.015"
>.080"	.020"

Table 5–17   Pad-to-Pad Spacing Depending on Pad Size

Pad Size	Minimum Spacing Requirements
<.030″	.008″
>.030″	.010–.012″

tor to cut the board to the center of the line. This is the way the manufacturer functions when creating a route for the board.

Slots

Slots are relatively the same as mounting holes but treated as a board edge. The same clearance values for the board-to-edge clearance (copper-to-edge clearance) apply. Check the table of manufacturer's capabilities to see what the minimum slot shall be. This is based on the minimum radius. The radius is one half of the router bit; thus the slot must be twice the radius (see Table 5–18). The radius width is also determined by the material thickness and based on hardness of the material.

Making Board Edge/Slot Clearance

A problem with many PCB software packages is the oversight in creating a rubber banding cutout/border that will build in clearances on appropriate layers. As shown in Figure 5–15, the inner layer copper will be exposed in a slot, the same as all cut edges of a board unless a clearance area is added to the design. Any copper layer in the design software should have a clearance area between the copper area and any routed edge. This is to prevent the router bit from pulling the copper from inside the board. A technique used in the design software is to use a board outline twice that of the desired clearance. Any slot or board edge should be drawn using this width of line. This line can then be copied to any negative plane layer, providing consistent clearance on all plane layers. A line width twice of the clearance is used in conjunction with a note specifying the manufacturer to route to the center of the board outline.

Table 5–18   Calculating Finished Pad Diameter

AR	+	Mfg AR	× 2	+ Hole	= Pad Diameter
.0025″	+	.009″	× 2		
.0025″	+	.009″	× 2		

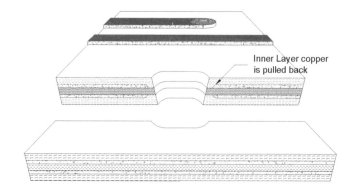

Figure 5–15   Slot in a multilayer board.

Note

A negative plane is a plane layer where any colored image is lacking copper and any clear area represents copper.

❏ *Add text to plane layer as to net name (GND, +5 V, etc.).*
❏ *Calculate board thickness and determine material availability. Attempt to use predefined or previously used combinations. Or, after design completion, save successful stack-up combinations.*
❏ *Add tooling holes, if appropriate.*

Tooling

In special cases tooling holes are necessary. Tooling holes are three or more holes placed at the board's extents used for routing special shape boards, such as round boards and some slots. If boards are beyond the normal square or rectangular boards, it is recommended to place tooling holes. Several different sizes are available, but a .125″ pin (.126″ hole) is used. Placing these holes to the extents of the boards provides stability from twisting during routing.

Fiducials

Fiducials come in several different sizes and shapes. Figure 5–16 shows the old-style fiducials or targets. These are used as targets for auto assembly, and are helpful for manufacturing. It is highly recommended to place at least three (or more) fiducials toward the extents of the board. This helps the manufacturer ensure alignment and is helpful for the designer to verify layer alignment.

Figure 5–16 Fiducial or target.

❑ *Add datums to all layers or overlay layer. This helps not only to verify alignment after completion but for manufacturing alignment.*

❑ *Board part number in copper on bottom side.*

❑ *Board revision (in copper or manually marking).*

❑ *Layer number (each layer, numbered by layer number, each offset).*

❑ *Assembly number (on silkscreen, topside).*

❑ *Assembly revision (leave blank area for manual marking).*

Initial checks

❑ *Check that power pins are connected correctly on one of each type of part.*

❑ *Check that plated-mounting holes are grounded when required.*

❑ *Complete placement location and prepare for routing.*

Manual Routing

❑ *Route the following types of nets first:*
   ❑ *Most difficult*
   ❑ *Most complex*
   ❑ *Tight fitting nets first*
   ❑ *Very high current (primarily external)*
   ❑ *Very high voltage (primarily internal)*
   ❑ *Sensitive*
   ❑ *Noisy*

❑ *Separate analog and digital.*

❑ *Route busses.*

Auto Routing

❑ *Manually route those items shown in "manual routing" first, if necessary.*

❑ *Define attributes that are commonly only to the auto router.*

❏ *Define/select "Routine," "Do" file, "Route" file, or "Strategy" file.*
❏ *After route completion,*
    ❏ *Manually clean up paths.*
    ❏ *Miter right angle corners.*
    ❏ *Run DRC/design rules to ensure clearances are met.*
    ❏ *Check annular ring.*

Component Placement and Routing Methodology

Components are placed primarily in order of function, easing routing require-ments. Some boards are designed for aesthetics, where all components are placed in the same orientation, and orientation is also a consideration for assembly ease and auto assembly. (See Chapter 2 for additional information.) Component align-ment also eases maintenance/servicing but may increase the complexity of rout-ing. If the board is of low or no maintenance, then component orientation can be overlooked, except for assembly issues.

Figures 5–17 through Figure 5–19 show several methods of routing. A board may use a combination of these styles, depending on the number of layers and component placement. Components may be placed, all in similar orientation, providing consistent pin 1 location, or they may be placed to accommodate rout-ing and reducing vias.

Alternate routing (Figure 5–19) is the simplest form alternating layers per direction. The top layer may be all horizontal lines and the bottom layer all verti-cal lines, or alternating each layer. Alternate routing is also used in some "noisy" designs to reduce parallelism and crosstalk.

The flip route method (Figure 5–18) is the same as alternate routing but oc-curs when the connecting pins are mirrored.

Note

Via reduction is important for high frequency design, reducing the number of layer transitions and reduces overall manufacturing cost calculated on number of holes drilled.

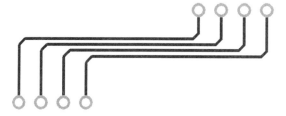

Figure 5–17 Simple bus type route.

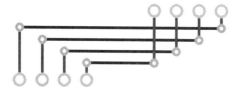

Figure 5–18   Flip route.

Determining Trace Width from Space Available

One controlling factor when determining trace width is space between pads on a connector. Regardless of current requirements, a trace must be routed. After the pads have been defined depending on assembly and serviceability and current requirements have been defined, the trace width is defined, as mentioned earlier in this chapter. If initial trace width will not fit, there are a few options:

- Change connector/components.
- Duplicate layer.
- Increase copper thickness.
- Change from internal layer (if internal) to external layer. (External traces may be smaller than internal.)
- Decrease clearance settings.

Remember that the trace weakness is the smallest width in a trace. If a trace was to break because of current, it will be at the narrowest point or the area with the smallest amount of copper. When tapering or necking up/down, it must be understood that the narrow part is the minimum width and the rest of the trace is a wider line than necessary.

Escape and Fan-Out

Component escape and fan-out are critical to designs and are commonly the areas that determine the trace width (see Figure 5–20). Escape from a component may even determine the number of layers required. A Pin Grid Array (PGA) is one of the most difficult components to fan out, and the designer may even have to have

Figure 5–19   Alternating route.

Figure 5–20   Escape clearance available.

one signal layer for each row of pin from the center. As shown in Figure 5–21, four rows may require three signal layers.

Attempt to route the first row on the bottom layer of the board. This will reduce/eliminate shorts or damage to the traces from soldering.

Wide Line Routing

Wide line routing is a term regarding a trace that is wider than required or the minimum trace width. This is recommended for traces in which

- Surges occur.
- There are unsubstantiated current requirements.
- Trace resistance is in question.
- The trace is constantly on.
- Temperature is high.

Wide line routing is used during routing for these reasons but may be reduced to the minimum when required. The advantage of wide line routing is heat dissipation for traces that have a constant current and have a high temperature expectancy.

Branch Circuits

Circuits do not always carry the same amount of current throughout. There is a source point and possibly several destinations. Each destination may not draw the same amount of current. Therefore, each branch of the same circuit does not necessarily need to be the same width (see Figure 5–22). This can only work if the software supports branch circuits (subnets) and the designer defines each branch

Figure 5–21   Escape from a PGA.

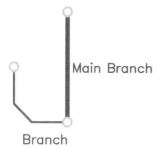

Main Branch

Branch

Figure 5–22   Branch circuits.

separately. Circuits such as ground and Vcc may have high current requirements, but the current cumulative. The cumulative current may be 5 Amps but is divided among five circuits at 1 Amp each. The traces may either fan out from one point with a trace width for only for 1 Amp, or a main branch can carry 5 Amps and then break out into separate 1 Amp traces.

Component Placement for Routing

Components such as connectors, switches, and lights sometimes have placement requirements and should be placed first. Component placement requirements are first determined by auto assembly if used and maintenance requirements. Discrete components such as resistors and capacitors should be placed in line with other components, or with each other to avoid path obstructions and to create routing channels (as shown in Figure 5–23).

Paths can also be predetermined and left clear of components. This allows the following:

- Separation of signals
- Matching lengths
- Reduced bends
- Reduced number of vias
- Reduced layers required

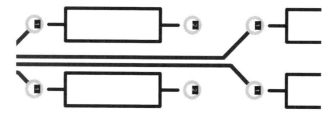

Figure 5–23   Inline component placement.

Form or Function

The constant battle in board design is of form or function. Application, board space, assembly, and a little philosophy drive this decision. Auto assembly dictates that the directions of components are placed primarily in the same direction. Manual assembly requires (or it is recommended) that the components be placed somewhat uniformly to avoid confusion during component population. Many companies like boards to be uniform and smooth in appearance. Designers want components to be placed in the best positions for easy routing. Board area requires that components be placed in a manner in which all components can be routed within the confined space using minimal layers. None of these are the only method, but all aspects are combined and considered during component placement. The compromise is placing components in a uniform direction but located to maximize route ability.

Primary Routing Layer

The primary routing layer isn't necessary but helps reduce layers and provides uniformity. SS boards will, of course, have only one side, and that side will be the primary routing layer.

A DS/ML board that is mostly thru-hole should have the bottom side as the primary routing layer. This is to reduce traces under components, provide better accessibility for cuts and jumpers, and alleviate heat and component noise.

A DS/ML board that is mostly surface mount requires the side with the surface mount components to be the primary routing layer. The surface mount components would require a via to connect the component to any other layer, thus making it more practical to use the same layer.

Primary Routing Direction

With an SS board there is no primary routing direction. All directions are required for routing.

With a DS/ML board, it is necessary to alternate directions on each layer, as shown in Table 5–20.

Table 5–20   Alternating Layer Directions

Layer	Direction
Layer 1	Horizontal
Layer 2	Vertical
Layer 3	Horizontal
Layer 4	Vertical

This accomplishes the following two things:

- Reduces parallelism between layers
- Provides horizontal and vertical routing paths

Many auto routers take this initial approach for routing of alternating layers whenever necessary and then removing unnecessary vias. Vias increase the space required between traces but usually shorten the routing path.

Single-Sided Route

A single-sided board is one of the most difficult because of the inability to use vias or change layers. Placement becomes critical, and all parts must be placed strategically with respect to route ability. Wire jumpers may be required to route the board if there isn't adequate space or there is no possible entrance to an area or to a component lead.

Before deciding to design a single-sided board, assembly time for wire jumpers against the additional cost of a double-sided board should be considered. In many cases where cost is critical, single-sided boards are a must.

Routing Bends/Miters

A bend or miter is an angle used in place of 90 degree corners. Traditionally the bend was necessary because the materials could stretch in the *x*- and/or *y*-axis, causing breaks in traces.

This stretching or swelling was due to the material properties and lack of humidity controls. With today's higher technology materials and better standards in material storage, stretching and swelling are rare, but the effects are proportional to the trace width. Just as a string is stretched, a wider string takes more than a smaller string of comparable material. Therefore, larger traces, those around and above .006″, will feel little effect from any swelling, but traces below that are more susceptible to changes in the material. Even if a designer only designs boards using larger traces, mitering traces is a good practice to help prevent etch pools and ensure spacing consistency throughout designs.

Etch pools are an event that occurs during the manufacturing process when a chemical etchant is used. Higher technology etch processes have all but eliminated this problem. Traces that have 90 degree corners tend to collect etchant, which will continue to eat away at the copper, resulting in a thinner trace width at the corner.

The spacing features presented with bends and miters allow traces to cut diagonally across a board, resulting in additional space or better use of the space available.

Many practices in design are not necessary for low technology boards but are a good practice for designs that may require such practices. If the same rules are used throughout all designs, then the designer will have a good handle on higher technology boards, reducing the learning curve.

Selecting the length and the spacing in a corner (Figure 5–24) comes down to philosophy and application. The easiest way is to start a bend at the first possible moment, which decreases available space (in some instances) while following as closely to other lines in the bend as possible (Figure 5–25).

This depends on application and space availability. Highly populated boards will have tighter, smaller bends than lightly populated boards. Other factors, such as length reduction and spacing for noise, become considerations in higher frequencies. Mitering can be based on trace width and/or grid spacing. Many times the grid spacing used is also based on trace width, so the following is a good rule of thumb:

- .025″ for traces under .012″
- .050″ bend for traces under .024″ to .014″
- .100″ bend for traces .049″ to .025″

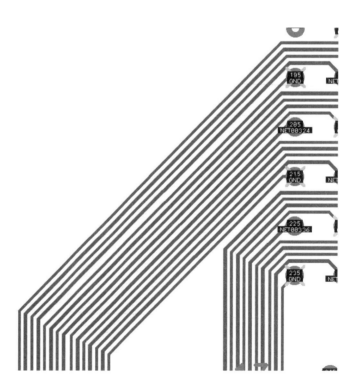

Figure 5–24   Tight bends.

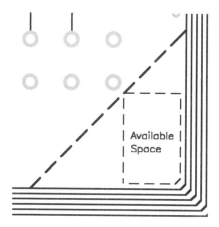

Figure 5–25   Set length bends.

Bus Routing

Many designs contain a group of connections that have similar beginning and end points. They may form a bus. In the schematic, the common thread is not always the same as in the PCB. The schematic's bus normally will be from one single component to individual and vice versa, or similar components, such as relays or several resistors that are of a common area or are going to a common area that are grouped together, creating a bus. Many times they are of a common signal or a common type of signal. Many/most of these buses will follow through to the PCB, and then additional lines that have either a common source, destination, or both will form, to create a bus.

Sensitive and protected signals are the first to be routed, then the bus lines. This is due to the number of lines that are contained, and they create a known amount of lines that will be contended with. Even other lines, not of the same group or area, may be included in the bus in some way to produce a clean path or a conduit of lines.

Caution must be used in the buses since parallel lines commonly cause crosstalk.

Noise, RF, EMF, Crosstalk, and Parallel Lines

There aren't clear guidelines between line length and spacing, along with the signal type. Weak or low voltage signals are susceptible to crosstalk, and "noisy" lines with many spikes and higher frequencies can inject noise into a parallel line that would not appear susceptible to noise. This event is known as saturation.

❏ *Change gates or parts.*

If a schematic was generated to help create the PCB, updating the schematic with PCB changes is critical! Any changes made should be reported back to the schematic. Most software packages allow what is known as "gate swapping" or allowing components that have identical parts/gates to be swapped with each other to provide a smoother and easier route.

Placement and Routing Interactivity

Form before function, or function before form? Defining the useable space, fixed placement components, as well as electrical/routing requirements determines the complexity of a board. No matter the spacing requirements, the electrical requirements are the controlling items. The design of the circuits determines the current and the voltage, two of the controlling items of the trace and space. The determination is then made to use thru-hole components or surface mount components from the amount of space available on the board and the area in the enclosure (if applicable). A consideration that is one of the largest factors and usually has the most impact is price.

Components are placed in their necessary position, such as potentiometers, displays, connectors, and mounting holes, and clearance area is determined.

Additional Markings

❏ *ESD symbol*
❏ *High voltage warning*

ESD labeling is a common practice for boards with components that are susceptible to damage from static discharge or physical contact that emits an electrical discharge. (Contact the ESDA or Electronics Static Discharge Association for additional information at http://www.esda.org.) High voltage warnings labels are a good practice and in some instances (specification and testing specific) are required.

After initial routing is complete and components are fixed, a design rule check, or DRC, is performed, verifying that electrical requirements are met.

❏ *Run DRC/design rules to ensure that clearances are met.*

Before polishing a design, it is good practice to ensure that all the rules and attributes are met. The same utility used in the schematic is used in the PCB. The DRC is a utility in most software packages that allows the user to check the design against the settings, attributes, and connections that were defined at the beginning. If clear values are not defined, then the program cannot check them, so it is important that those attributes required are defined. The most basic are as follows:

- Trace clearance (per design, net class, class-to-class and/or layer)
- Trace width (per design, net class, class-to-class and/or layer)
- Board edge
- Component clearance

If the numbers of errors are overwhelming, narrowing the check and troubleshooting one type at a time is recommended.

❏ *Relocate reference designators to their correct position/location/orientation.*

After the correct component locations are determined and most everything is in its place, the location of the reference designator (Ref Des) is next. Sometimes the location of the component is determined by the space required by the reference designator. Other times, when the designations are not critical, their locations are left to last and some are just left off. At the very least the designation should be placed under (or on top of) the component and used in the assembly drawing. Commonly an "all or none" approach is used to displaying the designators. It may be acceptable, if a troubleshooting manual will accompany the machine the board will be used in, that an outline of the board and the components with the designators displayed is the only source of locating components. See Chapter 8 for more information on reference designation locations.

❏ *Relabel/renumber reference designators.*

Relabeling or renumbering the Ref Des on a board depends on the application. It is strongly recommended that highly serviceable boards should be renumbered. This allows a logical order of components, so component locating is made easy.

The industry standard for renumbering is from the upper left-hand corner to the bottom right-hand. This is relative to reading a page in a book and provides logic and uniformity. This is not always the case and, in some applications, may be completely different.

Double-sided boards pose an additional problem. It is recommended that if all the designations are on the topside, then they be treated as if all the parts are on the top side, and if the designations are on the bottom side, along with the components, then they be treated separately and are renumbered from the upper left in regard to the orientation of the board when read.

Creating a Manufacturing/Fabrication Drawing

❏ *Copy border(s) to drawing layer or include border layer.*
❏ *Dimension the board in x- and y-dimensions.*

❏ *Hole to edge dimensions (this is used for registration verification for Gerber/drill loading).*
❏ *Dimension and tolerance of any cutouts under +/- .005″ tolerance.*
❏ *Board stack-up, to include the following:*
  ❏ *Layer number*
  ❏ *Layer type*
  ❏ *Layer thickness*
  ❏ *Layer tolerance*
  ❏ *Copper layer type*
  ❏ *Minimum trace width spacing per layer (special cases only)*
  ❏ *Overall board thickness*
  ❏ *Overall board tolerance (Conventional ±10%)*

Material Stack-Up

Toward the end of the design process, the numbers of layers have been determined. Along the way these values should be noted and a cross section placed in the fabrication drawing. Chapter 2 details the cross section display of the board with copper and dielectric thickness.

❏ *Drill legend, including*
  ❏ *Finished hole size*
  ❏ *Hole type (plated or nonplated)*
  ❏ *Hole tolerance (holes under .080″ +/-.003″; holes over .080″ +/- .005″; changes per technology)*
  ❏ *Symbol (correlates with fabrication drawing or Gerber export)*
❏ *Load or define fabrication notes, including*
  ❏ *Guidelines or specifications to follow unless otherwise noted (PC class [Quality] and type SS/DS or ML)*
  ❏ *Material used (core and Pre-Preg)*
❏ *Is copper thickness specified per table?*
❏ *Min. trace width and tolerance (+/-.003″ general .001″ tight)*
❏ *Min. clearance and tolerance (+/-.003″ general .001″ tight)*
❏ *Plating per same table (more plating, more plating in hole, increased MFG AR)*
❏ *Hole plating minimum of .0002 (usually external plating)*
❏ *Finish type: HASL or tin lead (check for availability)*
❏ *Hole to pad registration (no breakout allowed)*
❏ *Layers to layer 1 registration (+/-.002″)*

❏ *Overall scale tolerance (+/-.002 per inch. +/-.005" overall).*

❏ *Board size tolerance (+/-.005")*

❏ *Slot tolerance (+/-.003" to -.005")*

❏ *Beveling (if required)*

❏ *Electrical test and receipt of official test results. If required by P.O. or prototype only. Continuity of less than 5 ohms per inch. Test at 100 V.*

❏ *One or more of the following manufacturing markings (usually placed on the bottom side):*
  ❏ *Cage code (normally used by military contractors)*
  ❏ *Company logo (for identification if additional parts need to be ordered later in time)*
  ❏ *Date code (for board history)*
  ❏ *Lot code (for troubleshooting)*
  ❏ *Electrical test verification marking*

❏ *Twist and bow value (.010" def., .007" tight)*

❏ *Coupon or x-ray inspection for hole wall quality (one of the most important quality aspects of a board)*

❏ *Other*

As noted in Chapter 2, a fabrication drawing should, in two or less sheets, contain the following:

- Outline of the board with overall dimensions and a hole to board edge location. This helps align the board edge with the holes in the board and for board dimension checks.

- Fabrication notes—Define the standards for the manufacturer to follow and define other attributes above the standards and those not defined at all by the standards.

- Material stack-up with overall thickness and tolerance—The required overall thickness is a must. If not specified the manufacturer will define the layer and material thickness to their discretion. If layer thickness is not a concern, then this is an option.

- Drill table with tolerances and drill symbols in the board outline—This helps the manufacturer check that the drill sizes are correct and within the designer's requirements. Tolerances are different for different drill ranges.

Most of this information is not required but provides additional information that is recommended. Much of this information, such as the drill and board edge information, is used to compensate for the inconsistency between the software used by the designer and the manufacturer. It is highly recommended that this information be provided, but it might be replaced by simple notes directing the manufacturer to use its own standards or to follow the drill/Gerber data provided. IPC also specifies some tolerances and breakout values, which could reduce the need for drill tolerances.

Some manufacturers provide an "as is" service and view drill symbols and dimensions as redundant information and can provide a quicker turn and less expensive board with less information provided.

Documenting

❏ *Sheet/numbers of sheets*
❏ *Load or add information block specifying the following (this information may stay with the board until it is removed from the panel):*
  ❏ *Company name*
  ❏ *Company phone*
  ❏ *Layer name*
  ❏ *Layer number*
  ❏ *Part number*
  ❏ *Revision*
  ❏ *Sheet of sheets*

Application Company Specific Information

❏ *Add sheet revision block on first page (fabrication drawing).*
❏ *Add sheet revision section (border information).*
❏ *Update design information such as the following:*
  ❏ *Date (update every time this file is finished, changed, or modified)*
  ❏ *Designed by (designer name)*
  ❏ *Engineer (electrical engineer or the schematic's entry person)*
  ❏ *Checked by (QC, final, or engineer's name)*
❏ *Add sheet revision block on first page (Fabrication drawing).*

Check Plots (not required)

❏ *Print each layer w/o border to scale.*
❏ *Inspect for the following:*
  ❏ *Sheet layer numbers*
  ❏ *Datums*
  ❏ *ESD symbol*
  ❏ *HV note*

❑ *Tooling*
❑ *Pin 1 identification*
❑ *Mounting hole locations*
❑ *Board size and clearance*
❑ *Mechanical support*
❑ *Hardware clearance*
❑ *Stack-up thickness*

Approval

❑ *PCB approval from engineer.*
❑ *Implement any redlines.*
❑ *Generate netlist from schematic again.*
❑ *Run DRC again and run compare netlist.*

Output

❑ *Set up Gerber output files or set up database export.*
❑ *Export the following (in 274-X):*
  ❑ *All layer separately*
  ❑ *Required silk screen layers*
  ❑ *Top and bottom solder mask separately*
  ❑ *Fabrication drawing with symbols*
  ❑ *Drill file (in ASCII format, leading suppression)*
❑ *Load Gerbers in a CAM/CAD viewer and inspect for consistency with original design.*

SPECIFYING THE MANUFACTURING DO'S AND DON'TS

Manufacturers should never modify your files except to compensate for the manufacturer process, such as increasing trace width to account for etch-off. Manufacturers should contact the designer if the following occur:

- Board does not meet manufacturing requirements.
- Finished board will not meet IPC (or other standard) specifications for finished product.
- Board does not meet the fabrication drawing specifications.
- Board can't be built to the fabrication drawing specifications.

File Archive

❑ *May be done after receipt of board or test complete.*
❑ *Place files in restricted area and change file properties as Read only.*
❑ *If changes need to be made, change revision and start a new file using revision descriptor in the part number.*

TEMPLATES

A template is a collection of objects outlines and locations that are used consistently. The template usually is built around a board of a certain size, shape, and connector location or all. There are many different templates that may be made (e.g., templates for components, mechanical objects, text and title blocks). A template file is one that combines all of those necessary objects into one startup package. The amount of time that is saved from a template is enormous. Using a well-thought-out template may eliminate many steps in the design process.

These are the common elements of a PCB that may be encompassed in a template:

- Board size and shape.
- Board stack-up.
- Top assembly.
- Bottom assembly.
- Overlays. There may be more than one overlay that usually contains either objects placed on top of the board, the enclosure, or other particular objects that need to be accounted for.

There is such a large variety of objects on a PCB to choose from that it becomes essential to eliminate some of them and reduce the number to choose from. Any object that may be included in a template (software specific) that is consistent with each job should be included. For consistency, ensure that only objects that are used almost every time are included. Neglecting to do so can cause problems if the component is to be updated.

SUMMARY

The design process requires a multitude of information, such as cost, enclosure size (application), assembly, and maintenance requirements. If one of these aspects is overlooked a board may cost an excessive amount, or it must be redesigned to accommodate all requirements.

Using a design process helps to cover most of these aspects and requirements while tracking what tasks have been completed, including software requirements and features.

Seemingly simple items such as pad and trace sizes become critical in some events and the knowledge of these features is important so the designer may manipulate them according to the design's requirement.

Understanding technology and its relation to pricing is important when determining the cost of a design. The technology limits are also very important so a design may be kept in an expected price range and so the board is designed with realistic values and is able to be manufactured.

This chapter covers only the mechanical, electrical, and manufacturing requirements of the design. The aspects of EMF, RF, temperature radiation, component sensitivity, and high technology circuit requirements can fill a book, and many other books cover this topic.

6

Libraries, Components, and Data Sheets

This chapter will explain physical components, their relation to data sheets, their relation to PCB design software, and how they are represented by software. Such components are kept in the program's library so they can be placed and connected, allowing the designer to create intelligence behind a design. This chapter will help the designer

- Select, understand, and create component standards.
- Understand what a symbol is.
- Understand a pattern or footprint.
- Combine a symbol with a pattern.
- Read and understand a data sheet.

UNDERSTANDING COMPONENTS

Many terms used industry-wide look similar, but have different definitions, depending on the context. For a component, there is the software version and then there is the physical component. While the physical component is simply a group-

Related documentation: IPC-SM-782

ing of resistors, capacitors, ICs, and so on, the software component is relative to any object used in the software library, whether it is a resistor, capacitor, or simply a symbol representing power, wire, or a single hole. The software version consists of the physical component footprint and the schematic symbol, and possibly a three dimensional view.

The Two Halves of a Component

The two halves of a component refer to the *symbol* and the *pattern*. The schematic symbol is nothing more than a representation of the physical part, using a standard symbol or figure of the symbols function, and/or the function of the pins of the component. More than one schematic symbol may be used to represent a portion of a component. If a component is of multiple parts that are identical or is made of two or more individual sections, there may be several individual symbols as well. These symbols are commonly known as gates. A gate is traditionally a symbol representing an amplifier, buffer, logic function, but this term has also taken on multiple meanings. In PCB layout software, the definition of a gate is *any symbol (group or individual) representing a portion of a whole component.*

A good example is a 7400 NAND IC. This IC may be 14 pins with three pins per gate and two power pins. One pin is for the power and one for the ground or negative voltage, as shown in Figure 6–6. Each comparator may be broken into a gate. The power and ground may be made into an individual gate or may be connected behind the scenes as a power pin (a power pin requires a table or explanation on the schematic describing its connection).

The other example is of a relay, which is usually broken up into two or more parts, the first being the coil and the second being the contacts. The relay may even have multiple contact sets, which may be displayed as a group or individually. Figure 6–1 shows that the relay clearly has two separate groups: the coil and the contacts. This particular relay has two sets of contacts. They are shown together because they are mechanically connected and their states (open contact/

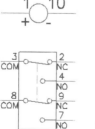

Figure 6–1　A relay.

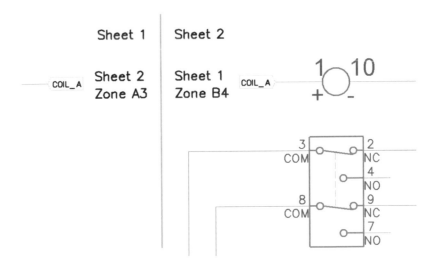

Figure 6–2   Single component using a sheet connector.

closed contacts) will change simultaneously. The dashed line connecting the two contact arms represents the mechanical connection.

Generically speaking, components should be broken down into the smallest parts within reason and built to be modular so they may be shown together and appear to be as one. Alternatively, the component may be shown as one single object and tags or sheet connectors may be used to connect a wire from one sheet or section to a component on another sheet or section, as shown in Figure 6–2.

Note

The component may be listed as a separate name/type to give the designer options of display; this will increase library size and complexity.

The complexity of different views is made more cumbersome by software. Some software doesn't allow alternate names or "titles" and simply uses the component's saved name. This doesn't allow the user to provide a short description of the component.

When a component name/type field is displayed, usually the actual name of the part is required to avoid confusion. Thus a component of a different style and the same name must be saved in different libraries. Ideally, software should allow a component to have different symbols and different patterns under one component name. Figure 6–3 shows the same component represented three different ways:

logic symbol†   **logic diagram (positive logic)**

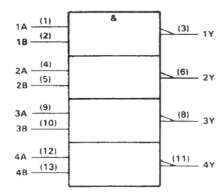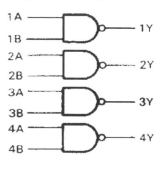

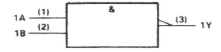

Figure 6–3   A four-gate component (7400 NAND).

- A block version with symbols on the pin representing the function
- The block gate view showing all the separate gates as one part
- A third view displaying only one gate of the component at a time

Much of these decisions are not made solely by the schematic/PCB designer but by an engineer or circuit designer or by the company's policy as to which type of schematics will be used (see Chapter 4).

COMPONENT CONSISTENCY

PCB design is made even more difficult because many components are created and designed using standards and measurement systems that do not conform or are not the same as the units system the designer is using.

Component Standards

There are several types of packages, footprints, and standards. Most of the following information is based on IPC's IPC-SM-782 or the latest revision. Most of these footprints are an adoption of industry-accepted formulas and standards. All footprints are derived from components that either conform to a package type or are new footprints/packages that do not, as of yet, have a standard.

Common Component Acronyms

- SMT—Surface mount technology.
- SMD—Surface mount design.
- JEDEC—Joint Electronic Device Engineering Council naming standards.
- SOIC—IC package with a SO (small outline).
- SO#—Small outline and number of pins. There are several assorted styles, so check your data sheet.
- ASIC—Application-specific integrated circuit.

COMPONENT SYMBOL TYPES

There are several formats of symbols that may be used, but the two common groups are the block style and discrete style. The block style is a symbol representing the complete component (Figure 6–4).

Note

A gate generally refers to a section of a logic-type component. This term has rolled over into defining a section of any component.

The other type or "gate" (shown in Figure 6–5) represents only a portion of the component. In some instances the gates may be swapped to use the best-positioned pins for smoother routing.

No particular type is correct. The choice depends on the types of schematics drawn and the types of connectors used. (This is covered in Chapter 4.)

The decision is made whether to use gates or sheet connectors/ports. Logic components are sometimes broken up into individual gates to show the logic functions. Other components are broken into individual sections to place on separate sheets depending on the function of the sheet. Sheet connectors or ports are useful but sometimes confusing when troubleshooting.

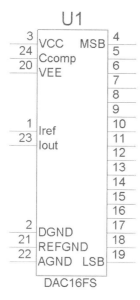

Figure 6–4   Block style.

A prime example is a relay. Relays can be broken into two definite parts: the coil and the contacts. A schematic may show a control section and a power section. The control section displays the logic controls and the coil of the relay. The power section shows higher currents or switching circuits and the contact portion of the relay.

Connectors are especially susceptible to breaking into gates, usually individual pins. This allows the connector to be broken into groups on different pages.

Components may be saved in more than one format. When saving components by part number and software only allows one style per component, saving style formats are difficult. An alternative is to use two different libraries, one for all block connectors and one for all gate/discrete connectors. For example:

• Connector_blk
• Connector_dis

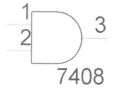

Figure 6–5   A simple discrete gate.

Table 6–1   Library Naming Scheme Matrix

	Use	Example
Component name will never be displayed with the component on either the board or schematic.	Use a suffix to denote package and symbol type.	7404_SM_Dis—surface mount and multiple symbols
Component name may be displayed with the component on either the board or schematic.	Use a separate library for each combination.	Logic_SM Logic_Dis Logic_SM_Dis

Use with programs that don't support multiple symbols/patterns types.

LIBRARY NAMING CONVENTION

SM components usually display the footprint *type* in the name of the footprint. This is critical when selecting components or when a particular part number/footprint is identified in a parts list. There are literally hundreds of footprints and package types. The footprint and the package type are not always identical.

Table 6–1 displays the library naming scheme for software that doesn't support multiple package styles.

MANUFACTURER-GENERIC VS. MANUFACTURER-SPECIFIC COMPONENTS

Many components, especially discrete components such as resistors and capacitors, are too generic and numerous to track a base component part number for each. Therefore, a pattern style or a base type may be created. Any/all parts that may be mounted to a PCB should have a pattern name, such as the SMT IC do. PLCC, SMD (surface mount dips), and PGAs are all standard footprints that a component may be referenced to. Manufacturers and electronics companies should define standard patterns to mate with all components. Until that time, designers create an assortment of part numbering schemes. Tables 6–2 and 6–3 use a common-sense scheme for assignment of discrete components.

Table 6–2   Capacitor Land Pattern Scheme for a Capacitor with a Radial Lead, Nonpolarized with a Pin Spacing of .200″, and a Lead Diameter of .030″

Capacitor	Radial/Axial	Polarized/NonPolarized	Pin spacing	Lead Diameter (Optional)
C	R	N	200	L030

Table 6–3   Resistor Land Pattern Scheme for a Resistor with Axial Lead with a Pin Spacing of .050", a Lead Diameter of .030", and a Wattage Rating of .025 Watts

Resistor	Radial/Axial	Pin spacing	Lead Diameter (Optional)	Wattage (Optional)
R	A	500	L030	W025

Commonly, resistors of the same wattage have the same lead diameter, and capacitors of the same voltage tend to have the same lead diameter. This is not always true, so lead diameter should be checked each time and lead diameter may be added to the scheme.

DECIPHERING A DATA SHEET AND MANUFACTURER'S STANDARDS—SMD

Whether a component is in hand or not, a data sheet (even mechanical components) is desired. This provides adequate information regarding an assortment of values. At times, especially for mechanical component, data sheets are unavailable and a manual notation of the item's properties must be created.

The initial engineer should attempt to archive all the data sheets used, in a central location available to others requiring this information. The data sheets should be stored in a location that is available to the engineer, PCB designer, and anyone involved in the project. This is an important step in reducing time and increasing throughput.

Many manufacturers use acronyms or part number extensions in their data sheets and part numbers to note specific attributes of their part. It is important to note these items or keep in some sort of record or reference sheet. Table 6–4 is a collection of common package styles and their appearance.

Package Styles by Manufacturer

The designer should keep a document with some of the component (manufacturer specific) nomenclature or part extensions (Table 6–5) and possibly, a table of common pin names or pin types (Table 6–6).

Not all parts that are made will be ICs. A majority of the components made will be resistors, capacitors, mechanical features, and other parts that are relevant to the types of boards the designer deals with. ICs are the most difficult components to create, so much of the focus will be on them.

Table 6–4   Standard Component Packages

PLASTIC		
Package Name	**Number of Pins**	
EBGA	352 pins	
LAMINATE CSP	16 pins, 24 pins, 28 pins, 48 pins, 128 pins	
LAMINATE TCSP	20 pins, 24 pins	
LAMINATE UCSP	24 pins	
LBGA	160 pins, 196 pins	
LLP	6 pins, 8 pins, 10 pins, 14 pins, 16 pins, 24 pins, 28 pins, 32 pins, 44 pins	
LQFP	32 pins, 44 pins, 48 pins, 52 pins, 64 pins, 80 pins, 100 pins, 144 pins, 176 pins	
MDIP	8 pins, 14 pins, 16 pins, 18 pins, 20 pins, 24 pins, 28 pins, 40 pins, 48 pins	
METAL QUAD	208 pins	
MICROSMD	4 pins, 5 pins, 6 pins, 8 pins, 9 pins, 10 pins, 14 pins, 20 pins	
MINI SOIC	8 pins, 10 pins	

Package Name	Number of Pins	
PBGA	352 pins	
PLCC	20 pins, 28 pins, 44 pins, 68 pins, 84 pins	
PQFP	44 pins, 64 pins, 80 pins, 100 pins, 128 pins, 132 pins, 160 pins, 208 pins	
PSOP	8 pins	
SC-70	5 pins, 6 pins	
SOIC NARROW	8 pins, 14 pins, 16 pins	
SOIC WIDE	14 pins, 16 pins, 20 pins, 24 pins, 28 pins	
SOP EIAJ TYPE II	24 pins	
SOT-223	4 pins, 5 pins	

(continued)

Table 6–4   Standard Component Packages (cont.)

Package Name	Number of Pins	
SOT-23	3 pins, 5 pins, 6 pins	
SSOP	16 pins, 48 pins	
SSOP-EIAJ	20 pins, 28 pins	
TO 220	3 pins, 5 pins, 7 pins, 9 pins, 11 pins, 15 pins	
TO 252	3 pins	
TO 263	3 pins, 5 pins, 7 pins, 9 pins, 11 pins	
TO 92	3 pins	
TQFP	48 pins, 64 pins, 80 pins, 100 pins	
TSSOP	14 pins, 16 pins, 20 pins, 24 pins, 28 pins, 48 pins, 56 pins	
TSSOP EXP PAD	20 pins, 28 pins	

HERMETIC

Package Name	Number of Pins	
CERDIP	8 pins, 16 pins, 18 pins, 20 pins, 24 pins, 28 pins	
CERPACK	10 pins, 14 pins, 16 pins, 20 pins, 24 pins, 28 pins, 48 pins	
CERQUAD	24 pins	
CQJB	44 pins	
CSOP	20 pins, 28 pins	
LCC	20 pins, 28 pins, 48 pins	
METAL QUAD	208 pins	
SIDEBRAZE	8 pins, 14 pins, 16 pins, 20 pins, 24 pins, 28 pins, 40 pins	
TO-3	2 pins, 3 pins, 4 pins	
TO-39	3 pins	
TO-4	4 pins	
TO-46	2 pins, 3 pins, 4 pins	
TO-5	6 pins, 8 pins, 10 pins	

The Data Sheet

For ease of use, one of the most common components will be used for an example of how to decipher a data sheet. The 7400 is a Quad NAND IC. Figures 6–6 and 6–7 show sample data sheets for a common IC.

As shown in Figures 6–6 and 6–7, combinations of views and representations should be displayed:

- The package type
- Number of pins
- Pin 1 location

Table 6–5 Package Styles by Manufacturers

Package Name	National	TI	Analog Devices	Motorola	Misc.
Metal leaded chip carrier, J bend	AA				
MQUAD	AW, MO				
MicroSMD	BP				
SIP	CA, C				
Sidebrazed, hermetic dip (metal seal)	DH, D, DA				
Leadless chip carrier (LCC)	EA, E				
Leaded hermetic quad packages	EL				
Ceramic flat pack (solder seal)	FA, F				
Dual leadless leadframe package (LLP)	LD				
Quad leadless leadframe package (LLP)	LQ				
Metal can, TO-5/39/46	HA, H				
CDIP	JA, J				
Metal can, steel TO-3	KA, K, KC, KS				
SO narrow	MA, M				
TSOP	MB, MBH, MBS, MDA, MDB				
SSOP (300 Mil wide)	ME				
SOT23	MF, M3, M5				
SC70	MG, M7				
TSSOP exposed pad	MH, MXA				
SOIC-mini	MM				
SOT223	MP				
SSOP (150 Mil wide)	MQ, MS, MSA, MSC, MQA, MEB, MEC, MED				
SSOP-EIAJ types II and III	MS				
TSSOP (4.4 mm and 6.1 mm wide)	MT				
SO wide	MW, WM				
MDIP	NA, N				
LTCC with castilations	SA				
LTCC with solder balls	SB				
LTCC with clip-on leads	SC				
LTCC with land contacts	SD				
Laminate CSP w/ land area contacts	SL, SLB				
FBGA (flex BGA)	SM, SLC				
TO220	TA, T				
TO202	TB, P, PA				
TO-252 (DPAK)	TD, DT				
TO220-ISO	TF				

Table 6–6   Pin type (by manufacturer)

Package Style	TI	Motorola	National	Fairchild	Misc.
Input pin	A				
Output pin	Y				
Power +voltage	$+V_{CC}$				
Power −voltage	$-V_{CC}$				
Ground pin	GND				

- Pin numbering layout
- Schematic symbol options
- Power pins
- Recommended voltages
- Operating temperatures

Note

The preceding items are in regard to PCB layout and schematic design/capture only. Other information is required for circuit design/engineering.

The package options for this IC are detailed as to what packages are available. (For ease the DIP package is used in (Figures 6–7 and Figure 6–8.)

After selecting the desired package, refer to the manufacturer's suffix to determine which picture/pinout is relative. For this example the N package has been selected.

Figure 6–9 helps to determine that it is a 14-pin DIP. (A common DIP pin spacing is .1″ from pin to pin and .3″ from row to row. This information is provided by the manufacturer's advance documentation in regard to the package types, or the package should come with your schematic capture software.)

Items such as input and output naming schemes are also common per manufacturer of the component rather than an industry-wide standard (Table 6–6).

The N type package also illustrates the notch used to mark pin 1 on conventional ICs and specifically shows the pin numbering in conjunction with the pin name.

The symbols may be of several different formats, and the one used may not necessarily look the same as that in the data sheet. The configuration of the symbol may follow a standard or may be determined by the designer. Different symbol options will be discussed later in this chapter.

• Package Options Include Plastic "Small
  Outline" Packages, Ceramic Chip Carriers
  and Flat Packages, and Plastic and Ceramic
  DIPs

• Dependable Texas Instruments Quality and
  Reliability

description

These devices contain four independent 2-input-
NAND gates.

The SN5400, SN54LS00, and SN54S00 are
characterized for operation over the full military
temperature range of −55°C to 125°C. The
SN7400, SN74LS00, and SN74S00 are
characterized for operation from 0°C to 70°C.

FUNCTION TABLE (each gate)

INPUTS		OUTPUT
A	B	Y
H	H	L
L	X	H
X	L	H

logic symbol†

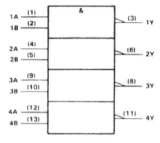

†This symbol is in accordance with ANSI/IEEE Std. 91-1984 and
IEC Publication 617-12.
Pin numbers shown are for D, J, and N packages.

SN5400 . . . J PACKAGE
SN54LS00, SN54S00 . . . J OR W PACKAGE
SN7400 . . . N PACKAGE
SN74LS00, SN74S00 . . . D OR N PACKAGE
(TOP VIEW)

```
    1A  [1    14]  Vcc
    1B  [2    13]  4B
    1Y  [3    12]  4A
    2A  [4    11]  4Y
    2B  [5    10]  3B
    2Y  [6     9]  3A
   GND  [7     8]  3Y
```

SN5400 . . . W PACKAGE
(TOP VIEW)

```
    1A  [1    14]  4Y
    1B  [2    13]  4B
    1Y  [3    12]  4A
   Vcc  [4    11]  GND
    2Y  [5    10]  3B
    2A  [6     9]  3A
    2B  [7     8]  3Y
```

SN54LS00, SN54S00 . . . FK PACKAGE
(TOP VIEW)

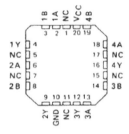

NC - No internal connection

logic diagram (positive logic)

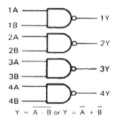

$Y = \overline{A \cdot B}$ or $Y = \overline{A} + \overline{B}$

Figure 6–6   7400 Quad NAND IC—Page 1.

recommended operating conditions

		SN5400			SN7400			UNIT
		MIN	NOM	MAX	MIN	NOM	MAX	
V_{CC}	Supply voltage	4.5	5	5.5	4.75	5	5.25	V
V_{IH}	High-level input voltage	2			2			V
V_{IL}	Low-level input voltage			0.8			0.8	V
I_{OH}	High-level output current			− 0.4			− 0.4	mA
I_{OL}	Low-level output current			16			16	mA
T_A	Operating free-air temperature	− 55		125	0		70	°C

electrical characteristics over recommended operating free-air temperature range (unless otherwise noted)

PARAMETER	TEST CONDITIONS †			SN5400			SN7400			UNIT
				MIN	TYP‡	MAX	MIN	TYP‡	MAX	
V_{IK}	V_{CC} = MIN,	I_I = − 12 mA				− 1.5			− 1.5	V
V_{OH}	V_{CC} = MIN,	V_{IL} = 0.8 V,	I_{OH} = − 0.4 mA	2.4	3.4		2.4	3.4		V
V_{OL}	V_{CC} = MIN,	V_{IH} = 2 V,	I_{OL} = 16 mA		0.2	0.4		0.2	0.4	V
I_I	V_{CC} = MAX,	V_I = 5.5 V				1			1	mA
I_{IH}	V_{CC} = MAX,	V_I = 2.4 V				40			40	µA
I_{IL}	V_{CC} = MAX,	V_I = 0.4 V				− 1.6			− 1.6	mA
I_{OS}§	V_{CC} = MAX			− 20		− 55	− 18		− 55	mA
I_{CCH}	V_{CC} = MAX,	V_I = 0 V			4	8		4	8	mA
I_{CCL}	V_{CC} = MAX,	V_I = 4.5 V			12	22		12	22	mA

† For conditions shown as MIN or MAX, use the appropriate value specified under recommended operating conditions.
‡ All typical values are at V_{CC} = 5 V, T_A = 25°C.
§ Not more than one output should be shorted at a time.

switching characteristics, V_{CC} = 5 V, T_A = 25°C (see note 2)

PARAMETER	FROM (INPUT)	TO (OUTPUT)	TEST CONDITIONS		MIN	TYP	MAX	UNIT
t_{PLH}	A or B	Y	R_L = 400 Ω,	C_L = 15 pF		11	22	ns
t_{PHL}						7	15	ns

Figure 6–7   7400 Quad NAND IC—Page 2.

● **Package Options Include Plastic "Small
Outline" Packages, Ceramic Chip Carriers
and Flat Packages, and Plastic and Ceramic
DIPs**

Figure 6–8   Package excerpt from data sheet.

SN5400 . . . J PACKAGE
SN54LS00, SN54S00 . . . J OR W PACKAGE
SN7400 . . . N PACKAGE
SN74LS00, SN74S00 . . . D OR N PACKAGE
(TOP VIEW)

```
        ┌──┬──┐
 1A □ 1 │  U  │ 14 □ VCC
 1B □ 2 │     │ 13 □ 4B
 1Y □ 3 │     │ 12 □ 4A
 2A □ 4 │     │ 11 □ 4Y
 2B □ 5 │     │ 10 □ 3B
 2Y □ 6 │     │  9 □ 3A
GND □ 7 │     │  8 □ 3Y
        └─────┘
```

Figure 6–9   Package excerpt from data sheet 2.

DRAWING THE COMPONENTS

Many companies utilize one single technician to create the component. Often this person is known as a librarian, and his or her duties are to create components while maintaining standards in naming conventions in addition to the physical appearance. The librarian may not make all the components, but coordination and template creation normally are the librarian's responsibility. All aspects of the components must be standardized to allow for smooth, consistent drawings. Items specifically controlled include symbol size, symbol/component font size, and grid pin spacing.

MULTIPLE ASPECTS OF THE SAME COMPONENT

Components may be represented in more than one way, there may be optional ways to display patterns (e.g., land patterns), and the pattern may need to be configured differently depending on the application. Again, this must be addressed in the software. The library part may be labeled by the pattern name generically or by the manufacturer's part number with the package extension, or the library may be labeled by package style.

In the future the software may label the part by the basic number and the designer may select from multiple land patterns.

Patterns

Patterns are listed first because they are the primary item in the creation of a component. The pattern represents the real-life component and the number of pins used, as well as options that may be contained in the pattern, such as wide and narrow versions of ICs and alternate patterns for resistors. Patterns can be made to accommodate several types of the same component, but a generic pattern name should be used. If software allows multiple patterns for a single part, multiple patterns should be generated so that each type of pattern may be defined and named individually. It is rare that a component with a specific name or part number will have multiple patterns. A multiple pattern part is more common with a custom part, which isn't a physical component at all but rather a grouping of holes, wires, or mechanical objects that can be configured differently. A multiple pattern may include the following:

- Mounting brackets
- Heat sinks
- Jumper holes

- Mounting holes
- Sockets
- Wiring points

Symbols

Symbols are created secondary to the pattern creation since all options of the component need to be addressed, such as mounting holes and alternate pads. Not all pins and holes of a component are required to be displayed. Here are a few that don't require display:

- Mounting holes that should be connected for ground
- Power pins ($+V_{cc}$, $-V_{cc}$, ground, etc.)
- Unused pins
- Pins always connected to ground

This is only in the case that the software allows automatic connection to nets with similar connection, or connection "behind the scenes." These pins are sometimes known as power pins or invisible pins. If these pins are not displayed with their respective gate, they should be noted in a table. All pins should be noted somewhere in the schematic. This eliminates confusion in diagnosis and troubleshooting.

Labeling Pin 1

Pin 1 is intricate to troubleshooting and the initial installation, especially with polarized components. There are several ways to denote pin 1:

- Square pad on pin 1
- The number 1 by the pin 1 pad
- For ICs a notch in the corner of the silk screen, or an arc type notch at the top of the IC silk screen

There are several more creative ways to do this, and some use a combination or all of these to ensure that pin 1 is clearly and easily found. Again this is determined by troubleshooting, maintenance, and assembly requirements. Silk screen notches and square pads should be used for assembly purposes primarily. Using text for pin 1 will clearly require more spacing around the component but is a clearer method of indicating pin 1.

Naming the Component

According to the designer's/company's requirements, the components may be saved as the actual part number, a generic version of the part number (e.g., 345-xxxx-D), or a purely generic description. This is usually based on the generality of the component, such as resistors and capacitors. Components by only one manufacturer should be saved as the manufacturer's part number.

SUMMARY

This chapter explained the two parts of a component, their relationship to each other, and how they are used in a schematic and a circuit board. Again, standardization in naming and displaying is the key to consistent components, allowing the engineer to convey clearly the connections of component to the designer. The creation of the component must take into account more than just the actual electronic function of the component and include the assembly of that part and pin identification for servicing of the board. In addition, naming schemes, documentation, and sorting of the components are essential when the library component count is in the thousands and quick identification and selection is critical.

7

Board Completion and Inspection

When boards are received from a manufacturer, an inspection should be done on all of the boards, including production copies. Prototype, or first-run, boards should be thoroughly inspected and the production copies should have at least a cursory inspection.

The manufacturer should have a quality inspection process in place to verify the final board in addition to an inspection at the completion of each process, but it is up to the purchaser to verify that the boards have been built to specification. There are some aspects that cannot be verified, such as the board material and the internal properties, but many other aspects can and should be verified.

Note

This chapter is brief but deserves individual attention because of its importance. The information provided should be used on every board. Additional tests should be preformed depending on the application, complexity, and requirements.

WHY INSPECT?

Not all specifications require inspection, but history will prove that a few minutes of inspection will avoid hours of troubleshooting, hours of conversing with the manufacturer, and return costs and assembly costs.

Incoming Board Inspection

❑ *Check the initial look of board. Note cleanliness and appearance.*
❑ *Check the mask, including*
   ❑ *Specified color*
   ❑ *Specified thickness*
   ❑ *Quality*
   ❑ *Blemishes*
   ❑ *Pitting*
❑ *Is plating adequate?*

Plating

Adequate plating is fairly difficult to measure. Current requirements are based on plating thickness, and if requirements are tight, then plating may become critical. The designer's simplest option is to mask the area near the board edge in front of or beside a pad. The unmasked area will act as a control point so when the pad height/thickness is measured, the board thickness may be subtracted from the pad height/thickness and divided by 1/2. This provides a measurement of thickness of the copper/plating/solder-flow combination. Regardless of plating thickness, the solder flow should be consistent. Table 7–1 shows theoretical measurements and must be adjusted to the finished thickness and specifications. This type of measurement is rarely used because of the small values and the irregularity of the solder flow.

❑ *Are holes centered in pad? (annular ring and alignment)*
❑ *Are hole sizes correct?*

Table 7–1   Plating and Solder-Flow Measurements for .062 and .093 Boards

Copper Thickness	Plating Thickness	Solder Flow	Total	Total × 2	Total × 2 + .062 board	Total × 2 + .093 board
.0007″	.0014″	.003″	.0051″	.0102″	.0722″	.1032″
.0014″	.0014″	.003″	.0058″	.0116″	.0736″	.1046″
.0028″	.0014″	.003″	.0072″	.0144″	.0764″	.1074″

Finished annular rings can be inspected by a cursory view or with an "eye loop" or a type of magnifying glass with an embedded scale and sample diameters. The finished annular ring should be no less than the specified annular ring, or the annular ring minus the manufacturer annular ring if adequate allowances were made.

Hole wall quality is essential but very difficult to test or verify. On all multilayer boards, a manufacturer should perform a cross-section test. Upon request, the manufacturer may provide a copy or verification that a test was preformed. A cursory test may be preformed to check continuity. A multimeter may be used to check resistance from one PLTH on a board to another PLTH that has an electrical connection. The resistance should be less than 1 ohm plus and additional 1 ohm per inch. A check between planes may be done to verify that there are no shorts or internal contamination. This value should be above 100 Mohm with no components mounted.

❑ *Are layers registered?*

Layer registration can be verified, including mask and silk screen, using datums internal to the board that are evident on every layer. These datums may be used by the manufacturer for the alignment process.

❑ *Perform an adhesion test.*

Place a piece of tape on plating/gold and remove. If plating or gold are removed with the tape, then the plating may be inadequate.

❑ *Does appearance match artwork?*

Verify that locations of text and lines are correct and all the holes are in their proper location (visually). Make sure that text thickness and line widths are correct and all extra markings are correct. Actual results are different from how they appear on paper, and this simple check ensures that the results are as desired. In production quantities, simple manufacturer mistakes (such as missing mounting holes) are not found until the board is assembled and there is an attempt to mount the board. Overlaying a 1:1 print on the board is a simple test for correctness. Use of film or a printable clear format can provide an extra measure of insurance.

❑ *Does the silk screen match artwork?*

Line width, text width, text size, and orientation should all match artwork. A cursory review of the board should determine if the silk screen matches the artwork (artwork being printouts or what is displayed on the computer screen).

❑ *Are overall board dimensions correct?*

If the tolerance for the board width or height is not specified, the default tolerance used by the manufacturer is +/-.005″. This must be confirmed with the manufacturer, or a generic note may be placed in the manufacturer notes stating that "all tolerances shall be less than +/-.005″ unless otherwise specified."

❑ *Is the board warped?*

Unless a perfectly flat surface is available, a counter or desktop is adequate to evaluate the flatness of a board. Generally the board should have little to no bow and the edges should rise very little. IPC has a standard test for measuring warpage, but a cursory test can provide adequate information in a commercial environment. Amount of bow is also a product of application. A manufacturer's cost for production and quality must be weighed to decide allowable amounts. Material warpage is a product of storage, and proper storage is costly. Some manufacturers trade storage costs with lower customer costs.

❑ *Is any copper showing on board edges?*

As noted in Chapter 5, there should be a copper clearance from the board edge unless required or necessary. This is to eliminate shorts and a tearing effect of the copper when the board is routed.

❑ *Trace width in tolerance.*

Trace width is another difficult item to measure. Without the aid of an eye loop, measuring trace width can only be due with a simple gauge. Trace tolerance and edge quality are also products of application and aren't always required to be tight and perfect. Rough trace edge is not extremely important until higher frequencies are reached.

❑ *Perform an electrical check on the following items:*
  ❑ *Several plated holes front to back (less than 2 ohms)*
  ❑ *Longest trace from one end to the other (less than 5 ohms per inch, or specified value)*
  ❑ *Resistance between planes of different nets for shorting (should be open)*
  ❑ *Test pads closest place for continuity (known opens should read open)*

A net or connections are a virtual short. There should be very little resistance, about 2 ohms or less per inch. Vias and thru-holes add resistance to a trace, but a poor thru-hole can show a much higher resistance and with the addition of heat (during assembly) may separate and read as an open. Several things can cause this:

- *Poor cleaning after drill*—During drilling, the heat can cause residue in the drilled hole and when the hole is plated and may not adhere well.
- *Poor pressing*—Inadequate heat and pressure during the pressing process may cause delamination later.
- *Excessive heat*—During other manufacturing and assembly process, excessive heat may be applied to the board.

Regardless of the reason, the result is usually the same: poor connection in the plated hole causing higher resistance. If a seemingly high resistance is present, normal soldering heat should be applied to the pad and then tested again.

SUMMARY

This chapter discussed the basics of finished board inspection. This process is important for new designers so they may see the results of their work and the realistic look of the finished design. The designer must review the completed board to determine if there are aspects of the board that need to be addressed for the next build or the next board designed.

The board inspection seems simple and possibly unimportant, but a board manufacturer isn't free of errors and will, from time to time, build a board incorrectly. It is important to capture these errors before the board is assembled. A hundred-dollar board may cost thousands after assembly when time and parts are calculated in. The short time it takes for this check will pay for itself when the first bad board is found.

8

Drawing an Assembly

This chapter was intentionally separated from Chapter 3, "Design for Assembly," to allow the designer to first understand how the boards are assembled and how the board should be designed. At that point, the designer has enough understanding to convey clearly the particulars of the board and what needs to be noted. The assembly drawing is also done after the board design is complete, so this chapter was placed respectively.

This chapter also appears brief in regard to the amount of information that could be detailed about assembly drawing, but this book is first and foremost about PCB design. To detail all aspects of the mechanical design task, a separate book would be necessary. There are already several books and standards dedicated to assembly drawings.

CREATING AN ASSEMBLY DRAWING

The decision must be made as to who creates the assembly drawing and if it will be drawn at all. Many companies employ mechanical draftspeople, and the assembly drawing is part of their duties. Others feel that since the designer did the rest of the board, he or she should also design the assembly drawing as well. Most of the components are there and only some additional information is required. Some companies divide the tasks between the two, having the designer input as much information as possible and then passing the design/drawing to the mechanical draftsperson, who will complete alternate views and add assorted mechanical components.

There are the single-person subcontractors or companies that *only* design boards. This chapter is directed to them. There are the assorted few that do no assembly drawing and provide enough on the silk screen to be suitable for assembly. Many designers also do the assembly and attempt this in the same software package with which the PCB is done, and one would expect stronger tools and more support for mechanical drafting. PCB designers have a long tradition of being CAD draftspeople who, for some reason (usually because of the knowledge of drawing programs), changed over to PCB design. As technology progresses, the distance between draftspeople and designers grows. Knowledge of basic drawing has given way to knowledge of electronics and components. As pay and reverence for the PCB designer grows, so will the trend of electronic engineers becoming designers or doing designs in addition to their normal task. As this progression occurs, knowledge of mechanical design and drafting decreases and the engineer either must learn to do more mechanical design or export portions of the design for a mechanical designer or draftsperson, who will to do the assembly drawing. The dilemma for small companies where the electronics engineer, PCB designer, and mechanical draftsperson becomes one in the same.

The following information should provide a simple hard and fast rule with which designers can provide standardization and increase simplicity:

- Determining the type of assembly drawing required has been covered in Chapter 3 but was geared toward the design aspect of the board. This chapter looks at the assembly type in regard to needs and requirements for service personnel and assembly technicians.
- Views will look at the best representation of the board for assembly technicians to understand how components are mounted to the board.
- Notes, leaders, and reference designators show the relationship among the board, parts list, and assembly notes.
- Revisions and versions take a look at different methods of configuring a board and how to label as such.

DETERMINING THE TYPE OF ASSEMBLY DRAWING REQUIRED

During the initial board design the assembly type and serviceability of the board should have been determined. Although some designers will decide one type of process for every board, others who design multiple types and technologies must

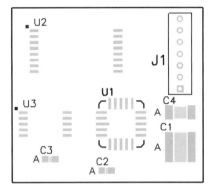

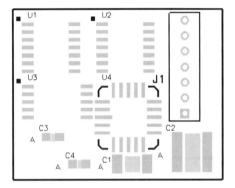

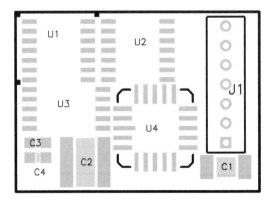

Figure 8–1   Reference designators outside the component area and large connector text.

Figure 8–2   Reference designators outside the component area.

Figure 8–3   Reference designators inside the component area.

utilize different types of assembly drawings according to applications and requirements:

- High serviceability and manually assembled boards need all the additional information available to the technician and assembler (Figure 8–1). This allows for easy reference when servicing and during assembly.
- If space is tight for board marking, then a board assembly and/or maintenance drawing should be created (Figure 8–2).
- Low serviceability and manually assembled boards require essentially the same as high serviceability and manual assembly boards, with the exception that the reference designators may be hidden (Figure 8–3).

ASSEMBLY VIEWS

The assembly drawing need only display a minimum of two sides. The side view displays the height of the components to verify the components and positioning. The initial view should be of the top of the board, displaying the outline of the components (or a solid graphic of the component if necessary) and the reference designator, creating a realistic view of the completed board. The hard and fast rule is to display all aspects necessary to convey where and how the board is to be assembled. Small aspects, such as connector pin insertion, keying orientation, and switch positions, should be enlarged or exploded along with assembly notes to express how the board is assembled. This upper-level assembly should display only those components hard mounted to the board. These are objects that are soldered, snapped on, or screwed to the board. When the board is mounted to another object, the board will become a subassembly of the next higher assembly.

As shown in Figure 8–4, the entire assembly drawing may be shown on one single page. The drawing consists of the following:

- Title block
- Top view of board
- Side view of board
- Assembly notes
- Find numbers (for parts with no reference designators)
- Note find numbers (reference notes to find numbers or parts)
- Optional reference designators table
- Optional wire connection table
- Optional ESD note

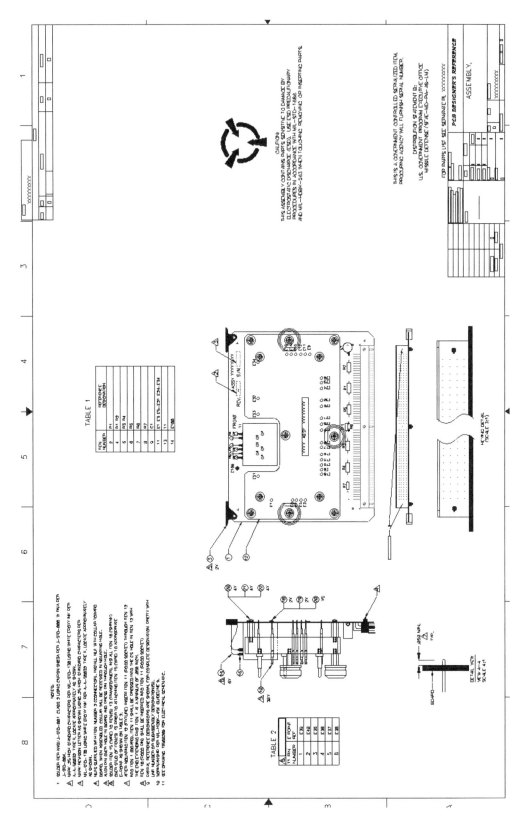

Figure 8-4 Sample assembly drawing (single page).

Merging the Silk Screen

Much of the information included in the assembly drawing is provided in the silk screen, as shown in Figure 8–4 (if one is used). As noted in Chapter 6, all the aspects of a component should be included with the component. This includes the overall assembly and anything required in the assembly drawing. A majority of the component information should be included in the silk screen. There may be spaces where lines are broken because of pad clearance, board edge, or portions of the connector that are not on the board. These can be inserted in the assembly layer of the component. This provides a seamless outline of the connectors and component in the assembly drawing.

Assembly Drawing Checklist

The top view of the assembly drawing may consist of the component outlines and reference designators (or the silk screen) and any additional markings or hardware, such as

- Screw
- Nuts
- Wire
- Shrink tubing
- Heat sinks
- Mounting brackets
- Mounting hardware

❏ *Show the top view of the board, including*
   ❏ *Reference designators*
   ❏ *Component outline (both may be imported from the silk screen)*
   ❏ *Leader to any component w/o a reference designator.*

These are only a few of the elements that make up the assembly drawing. Again, this is all dependent on the application and the servicing of the board. The intent is to include everything that is on the board, including subassemblies (as shown in Figure 8–5). Installation and exploded views are also possibilities (Figure 8–6). These are for advanced boards that require components that cover other components, obscuring the view from almost any angle.

❏ *Side view of board: Show additional views of the board, if necessary, to show the assembly clearly.*

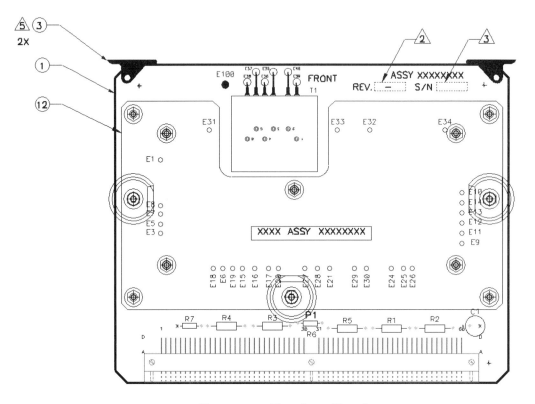

Figure 8–5 Top view of board.

The side view of the board (Figure 8–7) provides an aspect of the board that is not clear from the top-down view. Component height and orientation are some of the aspects to consider. Most component or hardware that requires assembly from both sides of the board is noted on the side view. Such items include

- Mounting hardware
- Standoffs
- Items specifically detailed in the side view.
- 90-degree connectors

Figure 8–6 Front view of board showing keying detail.

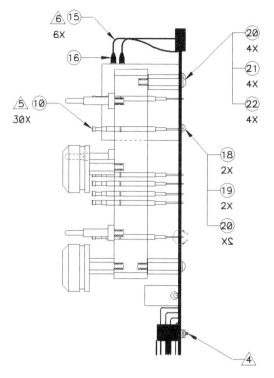

Figure 8–7   Side view of board.

- Wire detail
- Overall assemblies

❑ *Show this assembly separate from its enclosure (enclosure is next higher assembly)*
❑ *Load/Place assembly notes, such as:*
   ❑ *Soldering type*
   ❑ *Install first*
   ❑ *Part list information*
   ❑ *From/to information*
   ❑ *Pressing or flaring of mounting parts*
   ❑ *Other additional requirements*

Assembly Notes

Assembly notes (Figure 8–8) should answer all questions, including soldering standard, workmanship standards (breaking sharp edges, removal of excess lead length, etc.), and even the company's written standards. Designers must either draw from

NOTES:

1   SOLDER PER ANSI/J−STD−001, CLASS 2 USING CMPSN SN63A PER J−STD−006, W, RMA PER
    J−STD−004.

⚠2 MARK .25 HIGH STANDARD CHARACTERS PER MIL−STD−130 USING WHITE EPOXY INK PER
    A−A−56032, TYPE II. LOCATE APPROXIMATELY AS SHOWN.

⚠3 MARK REVISION LETTER AS SHOWN USING .25 HIGH STANDARD CHARACTERS PER
    MIL−STD−130 USING WHITE EPOXY INK PER A−A−56032, TYPE II. LOCATE APPROXIMATELY
    AS SHOWN.

⚠4 NUTS SUPPLIED WITH ITEM NUMBER 2 (CONNECTOR). INSTALL NUT WITH COLLAR TOWARD
    BOARD. WHEN ASSEMBLED, COLLAR WILL BE RECESSED IN MOUNTING HOLE.

⚠5 ALIGN PIN OVER HOLE IN BOARD AND PRESS PIN THROUGH HOLE.

⚠6 SOLDER ITEM 15 (WIRE) TO ITEM(S) 13 (TRANSFORMER). INSTALL ITEM 16 (SHRINK)
    OVER STUD OF ITEM(S) 13 PRIOR TO ATTACHING ITEM 15 (WIRE) TO APPROPRIATE
    E−POINT AS SHOWN ON TABLE 2.

⚠7 AFTER MOUNTING ITEM 12 (FIXTURE) INSERT ITEM 11 (POGO SOCKET) THROUGH ITEM 12
    INTO ITEM 1 (BOARD). ITEM 11 SHALL BE PRESSED INTO THE CS HOLE IN ITEM 12 WITH
    THE END EXTENDING PAST ITEM 1 AT A MINIMUM OF .050 INCH.

⚠8 ITEM 10 (POGO PIN) SHALL BE INSERTED INTO ITEM 11 (POGO SOCKET)

9   PARTIAL REFERENCE DESIGNATORS ARE SHOWN. FOR COMPLETE DESIGNATION, PREFIX WITH
    UNIT NUMBER OR SUBASSEMBLY DESIGNATION(S).

10 SEE DRAWING XXXXXXXX FOR ELECTRICAL SCHEMATIC.

11 UNLESS OTHERWISE SPECIFIED THE FINISHED ASSEMBLY SHALL BE IN ACCORDANCE WITH
    IPC−A−610.

Figure 8–8   Assembly notes.

their own experience in assembly or consult the assembly personnel if there are assembly process questions. After the assembly process is understood, the designer must imagine assembling the product and answer any questions that arise.

Since many objects on a board are mechanical in nature, or because the board is only part of an overall mechanical assembly, some designers/companies feel that board assembly should be completed by a mechanical designer.

If a board assembly is contracted out (possibly assembled by several different companies over time), even the smallest details should be addressed. This may include assembly order, step-by-step images with accompanying notes, marking colors, and even the "amount of bend" in metal tabs used for shields, covers, and enclosures.

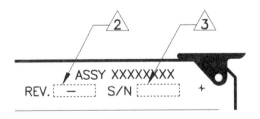

Figure 8–9   Stamping of assembly revision or version.

CAUTION:
THIS ASSEMBLY CONTAINS PARTS SENSITIVE TO DAMAGE BY
ELECTROSTATIC DISCHARGE (ESD). USE ESD PRECAUTIONARY
PROCEDURES IN ACCORDANCE WITH MIL—STD—1686
AND MIL—HDBK—263 WHEN TOUCHING, REMOVING, OR INSERTING PARTS.

Figure 8–10   ESD note.

Note

Notes and find numbers for PCB assembly should follow mechanical drawing standards.

❏ *Dimensions for conformal coating.*
❏ *Assembly Revision placement.*

The stamping or marking of the assembly version or revision is optional and is best left off of the silk screen and placed according to the configuration of the assembly. This allows the designer to make one board with several assembly configurations (Figure 8–9).

❏ *ESD symbol and note (if necessary).*

If applicable, ESD logos and notes (as shown in Figure 8–10) should be placed to ensure that assembly personnel are aware of the precautions that should be followed during assembly.

ASSEMBLY DRAWING FINAL NOTE

There is no single way to draw an assembly drawing. It depends mostly on the application and assembly requirements. As noted, there are many different ways to draw an assembly drawing, using several different programs. One way may not be suitable for a company, and multiple formats may work, especially in the case of a contractor. The requirements then become the customer's requirements, and the designer must be prepared to draw several different ways and in several different formats. There are several drawing standards that may be used to determine the formats.

SUMMARY

Many elements in an assembly drawing are based on the requirements of the assembler and possibly the needs of a service technician who will use this drawing for troubleshooting purposes.

The assembly drawing should clearly convey the assembly of the board form any angle required to display intent clearly. All hardware for the board assembly should be included along with notes on how parts should be assembled or in what order.

Additional notes for the finished assembly may be necessary and should conform to an existing, proven standard to reduce the amount of detailed information.

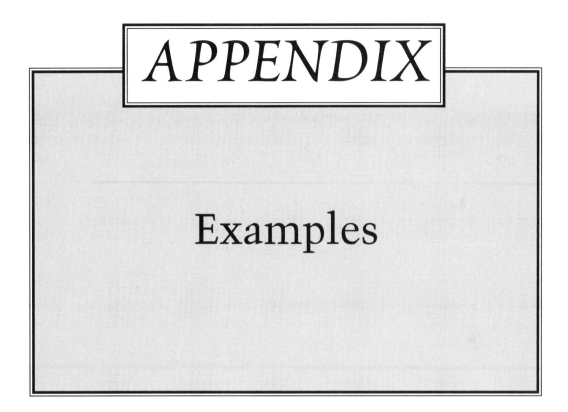

APPENDIX

Examples

SCHEMATICS

The following figures are examples of schematics and schematic styles/formats.

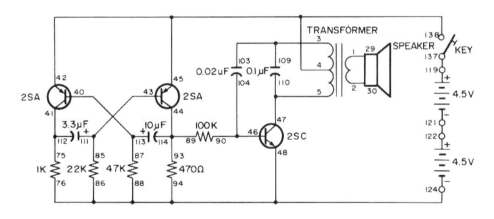

Figure B.1.   Alarm Circuit

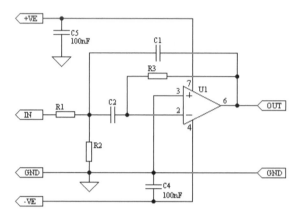

Figure B.2.   Filter Circuit

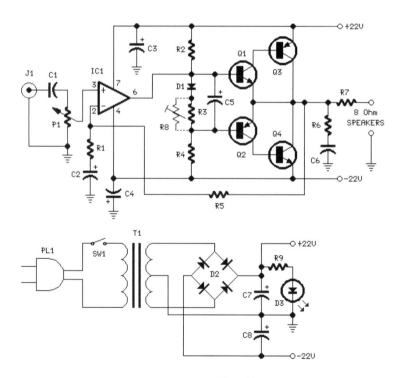

Figure B.3.   Amplifier Circuits

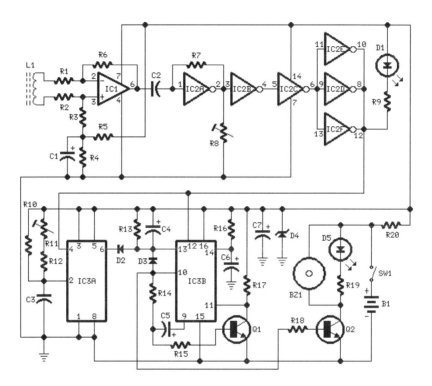

Figure B.4. Speed Limit Sensor

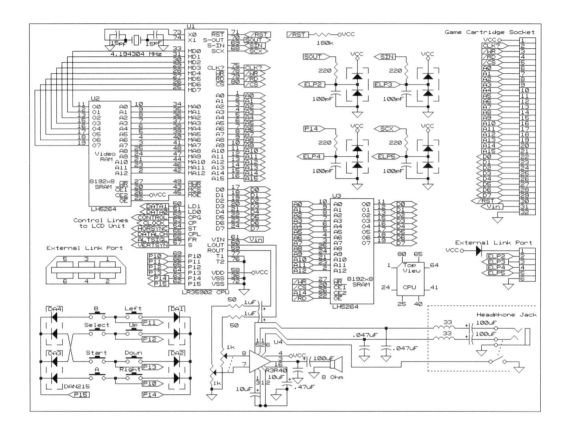

Figure B.5. Electronic Games

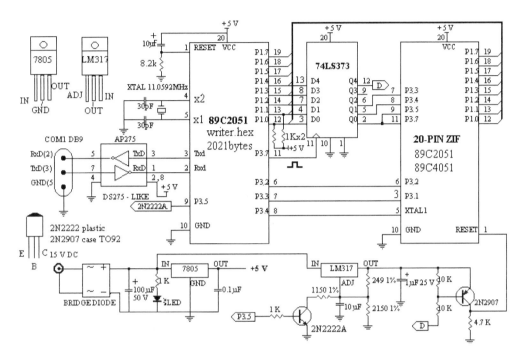

Figure B.6.   Computer Boards

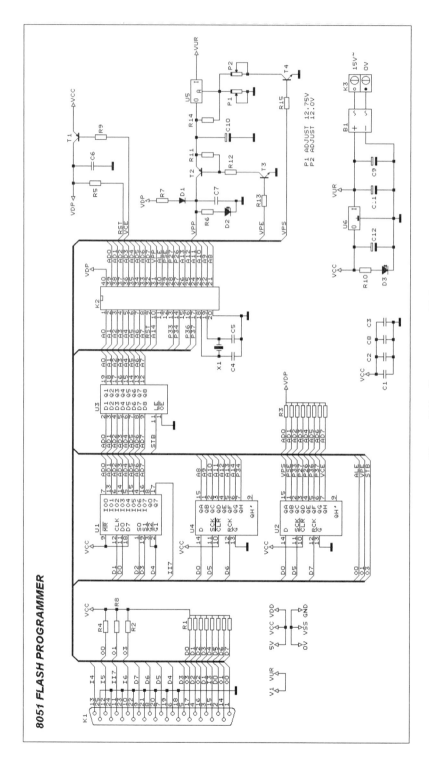

Figure B.7. Processors and Memory

176

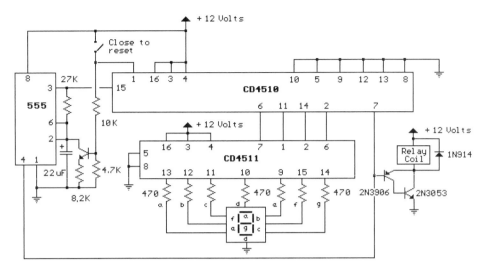

Figure B.8. Readout Countdown Timer

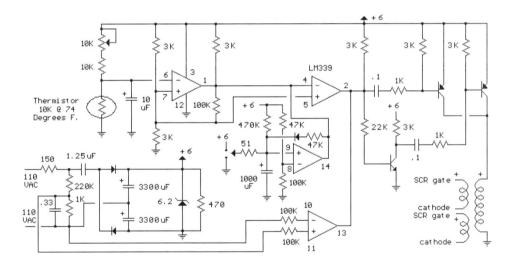

Figure B.9. Heater Controller

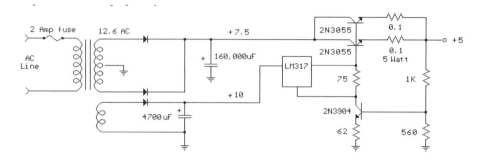

Figure B.10.   Video Amplifier

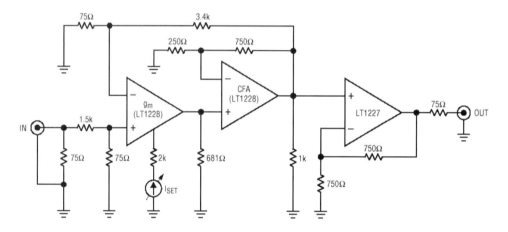

Figure B.11.   Intercom

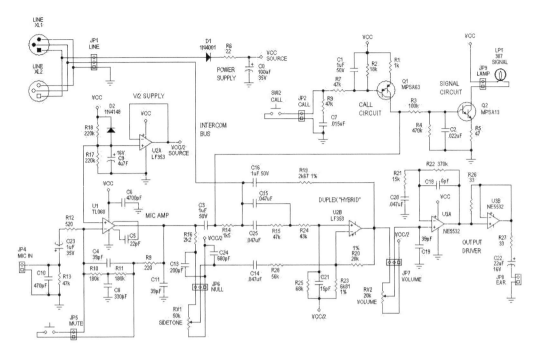

Figure B.12. High Current Power Supply

PCBs

The following figures are examples of different styles of PCBs. All are functional but are designed around the application, environmental conditions, UL requirements, or designer's standards.

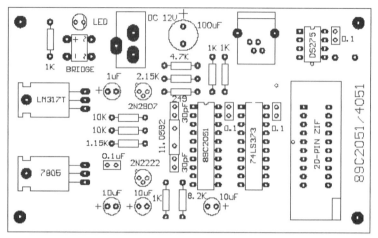

Figure B.13. Misc. 1

Figure B.14. PCB with Audio

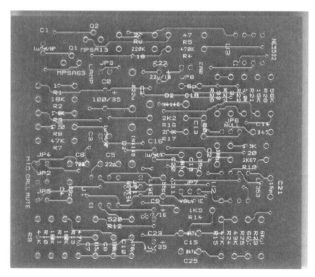

Figure B.15. Intercom PCB

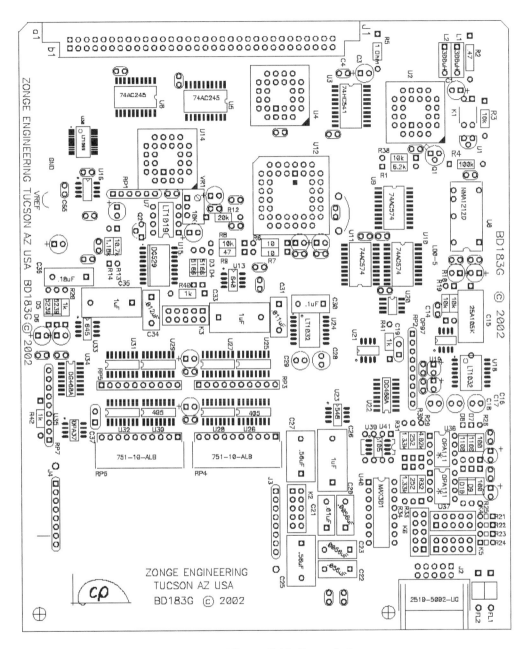

Figure B.16. Example 1

PCB Manufacturing Terms

These terms are those used by PCB manufacturers. Some terms have a standard definition, but when used as a manufacturing term, take on a different meaning. These glossaries are shown in their entirety from the contributor.

Term	Definition
Activating	A treatment that renders nonconductive material receptive to electroless deposition. Also called seeding, catalyzing, and sensitizing.
Additive Process	Any process in printed circuit board manufacturing where the circuit pattern is produced by the addition of metal.
Analytical Services Lab	Performs various tests such as plating thickness, inner layer connections to hole walls, photos or x-rays of circuit boards when required.
Annular Ring	The conductive material surrounding a hole.
Artwork Master	An accurately scaled (usually 1:1) pattern, which is used to produce the production master.

PCB manufacturing terms courtesy Merix Corp. (*www.MERIX.com*)

Term	Definition
Aspect Ratio	The ratio of the circuit board thickness to the smallest drilled hole diameter.
B-Stage Material	Sheet material (fiberglass cloth) impregnated with a resin cured to an intermediate stage (B-stage resin). Pre-preg is the preferred term.
Back Planes and Panels	Interconnection panels onto which printed circuits, other panels, or integrated circuit packages can be plugged or mounted. Typical thickness is 0.125"–0.300."
Barrel	The cylinder formed by plating a drilled hole.
Base Copper	Copper foil provided in sheet form or clad to one or both sides of piece of laminate used as either internal or external layers of a circuit board.
Base Laminate	The dielectric material upon which the conductive pattern may be formed. The base material may be rigid or flexible.
Base Material	*See Base Laminate.*
Bed-of-Nails Technique	A method of testing printed circuit boards that employs a test figure mounting an array of contact pins configured so as to engage plated thru-holes on the board.
Bleeding	A condition in which a plated hole discharges process material or solution from crevices or voids.
Blind Via Hole	A plated-through hole connecting an outer layer to one or more internal conductor layers of a multilayer printed board but not extending fully through all of the layers of base material of the board.
Blister	A localized swelling and separation between any of the layers of a laminated base material, or between base material and conductive foil. It is a form of delamination.
Blow Hole	A solder joint void caused by out-gassing of process solutions during thermal cycling.
Bond Strength	The force per unit area required to separate two adjacent layers of a board when applied perpendicular to the board surface. *See Peel Strength.*
Breakdown Voltage	The voltage at which an insulator or dielectric ruptures, or at which ionization and conduction take place in a gas or vapor.
Bridging, Electrical	The formation of a conductive path between two insulated conductors such as adjacent traces on a circuit board.
Buildability Meeting	Team meeting to review customer designs against manufacturing process capabilities. Used to identify possible failure modes prior to fabrication.
C-Stage	The condition of a resin polymer when it is in the fully cured, cross-linked solid state, with high molecular weight.
Center-to-Center Spacing	The nominal distance between the centers of adjacent features or traces on any layer of a printed circuit board. Also known as "pitch."
Chamfer	A corner that has been rounded to eliminate an otherwise sharp edge.
Characteristic Impedance	A compound measurement of the resistance, inductance, conductance and capacitance of a transmission line expressed in ohms. In printed wiring its value de-

Term	Definition
	pends on the width and thickness of the conductor, the distance from the conductor to ground plane(s), and the dielectric constant of the insulating media.
Chase	The aluminum frame used in screening inks onto the board surface.
Circuitry Layer	The layer of a PCB containing conductors, including ground and voltage planes.
Clad or **Cladding**	A thin layer or sheet of copper foil, which is bonded to a laminate core to form the base material for printed circuits. *See Base Copper.*
Clearance Hole	A hole in the conductive pattern larger than, but concentric with, a hole in the printed board base material.
Coefficient of Expansion, Thermal	The fractional change in dimension of a material for a unit change in temperature.
Component Hole	A hole used for the attachment and electrical connection of component terminations, including pins and wires, to the printed circuit board.
Component Side	That side of the printed circuit board on which most of the components will be mounted.
Conductive Pattern	The configuration or design of the conductive material on the base laminate through which electrical energy passes. Includes conductors, lands, and through connections.
Conductor	A thin conductive area on a PCB surface or internal layer usually composed of lands (to which component leads are connected) and paths (traces).
Conductor Base Width	The conductor width at the plane of the surface of the base material. *See Conductor Width.*
Conductor-to-Hole Spacing	The distance between the edge of a conductor and the edge of hole.
Conductor Thickness	The thickness of the trace or land including all metallic coatings.
Conductor Width	The observable width of the pertinent conductor at any point chosen at random on the printed circuit board.
Contaminant	An impurity or foreign substance whose presence on printed wiring assemblies could electrically, chemically, or galvanically corrode the system.
Continuity	An uninterrupted flow of electrical current in a circuit.
Coordinate Tolerancing	A method of tolerancing hole locations in which the tolerance is applied directly to linear and angular dimensions, usually forming a rectangular area of allowable variation. *Also see Positional Limitation Tolerancing* and *True Position Tolerance.*
Copper Foil	*See Base Copper* and *Clad or Cladding.*
Core Group	Daily operational meeting held on each shift to communicate current status of plant in terms of producing and delivering high quality, cost-effective circuit boards to customers on time. The group identifies problems that may prevent reaching daily/weekly goals and plans resolution of those problems.

Term	Definition
Cosmetic Defect	A defect such as a slight change in its usual color that doesn't affect a board's functionality.
Cover Lay, Cover Layer, Cover Coat	Outer layer(s) of insulating material applied over the conductive pattern on the surface of a printed circuit board.
Crazing	A condition existing in the base material in the form of connected white spots or "crosses" on or below the surface of the base material, reflecting the separation of fibers in the glass cloth and resin material.
Current Carrying Capacity	The maximum current which can be carried continuously, under specified conditions, by a conductor without causing degradation of electrical or mechanical properties of the printed circuit board.
Datum Reference	A defined point, line, or plane used to locate the pattern or layer for manufacturing, inspection, or for both purposes.
Deburring	Process of removing traces of base copper material that remain around holes after board drilling.
Defect	Any deviation from the normally accepted characteristics of a product or component. *Also see Major Defect and Minor Defect.*
Definition	The accuracy of pattern edges in a printed circuit relative to the master pattern.
Delamination	A separation between any of the layers of a base material or between the laminate and the conductive foil, or both.
Desmear	Removal of epoxy smear (melted resin) and drilling debris from a hole wall.
Develop	An imaging operation in which unpolymerized (unexposed) photo-resist is dissolved or washed away to produce a copper board with a photo-resist pattern for etching or plating.
Dewetting	A condition that occurs when molten solder has coated a surface and then recedes, leaving irregularly shaped globules of solder separated by areas covered with a thin solder film; base metal is not exposed.
Dielectric	An insulating medium, which occupies the region between two or more conductors.
Dielectric Constant	The ratio of permittivity of the material to that of a vacuum (referred to as relative permittivity).
Digitizing	Any method of converting feature locations on a flat plane to digital representation in x-y coordinates.
Dimensional Stability	A measure of dimensional change caused by factors such as temperature, humidity, chemical treatment, age or stress; usually expressed as units/unit.
Dimensioned Hole	A hole in a printed circuit board where the means of determining location is x-y coordinate values not necessarily coinciding with the stated grid.
Double-Sided Board	A circuit board with conductive patterns on both sides.
Drills, Circuit Board	Solid carbide cutting tools with four facet points and two helical flutes designed specifically for the fast removal of chips in extremely abrasive materials.

Term	Definition
Dry Film Resists	Coating material in the form of laminated photosensitive sheets specifically designed for use in the manufacture of printed circuit boards and chemically machined parts. They are resistant to various electroplating and etching processes.
Edge Bevel	A bevel operation performed on edge connectors to improve their wear and ease of installation.
Edge-Board Connector	A connector designed specifically for making removable and reliable interconnection between the edge board contacts on the edge of a printed board and external wiring.
Edge Dip Solderability Test	A solderability test performed by taking a specially prepared specimen, fluxing it with a nonactivated rosin flux, and then immersing it into a pot of molten solder at a predetermined rate of immersion for a predetermined dwell time, and then withdrawing it at a predetermined rate.
Electroless Plating/ Electroless Deposition	The disposition of metal from an auto catalytic plating solution without application of electrical current. Short for "electrodeless." This process is required to plate the nonconductive hole walls in order that they may be subsequently electroplated. Also called "PTH."
Electroplating	(1) The electrodeposition of a metal coating on a conductive object. The object to be plated is placed in an electrolyte and connected to one terminal of a DC voltage source. The metal to be deposited is similarly immersed and connected to the other terminal. Ions of the metal provide transfer to metal as they make up the current flow between the electrodes. (2) The electrolytic process used to deposit a metal on a desired object by placing the object at one electrical polarity and passing a current through a chemical solution to another electrode. The metal is plated from either the solution or the other electrode.
Entrapment	The damaging admission and trapping of air, flux, and/or fumes; it is caused by contamination and plating.
Epoxy Smear	Epoxy resin that has been deposited onto the surface or edges of the conductive inner layer pattern during drilling. Also called Resin Smear.
Etch	Chemical removal of metal (copper) to achieve a desired circuit pattern.
Etch Factor	The ratio of the depth of etch (conductor thickness) to the amount of lateral etch (undercut).
Etchback	The controlled removal of all components of base material (glass and resin) by a chemical process on the sidewall of holes in order to expose additional internal conductor areas.
Fiducial	Etched features or drilled hole used for optical alignment during assembly operations.
Film Artwork	A positive or negative piece of film containing a circuit, solder mask, or nomenclature pattern.
First Article	A sample part or assembly manufactured prior to the start of production for the purpose of assuring that the manufacturer is capable of manufacturing a product that will meet specified requirements.

Term	Definition
Fixture	A device that enables interfacing a printed circuit board with a spring-contact probe test pattern.
Flat	A standard size sheet of laminate material, which is processed into one or more circuit boards.
Flux	A substance used to promote or facilitate fusion such as a material used to remove oxides from surfaces to be joined by soldering or welding.
Fused Coating	A metallic coating (usually tin or solder alloy) that has been melted and solidified, forming a metallurgical bond to the base material.
Glass Transition Temperature	The temperature at which an amorphous polymer (or the amorphous regions in a partially crystalline polymer) changes from a hard and relatively brittle condition to a viscous or rubbery condition. When this transition occurs, many physical properties undergo significant changes. Some of those properties are hardness, brittleness, coefficient of thermal expansion, and specific heat.
Grid	An orthogonal network of two sets of parallel, equidistant lines used for locating points on a printed circuit board.
Ground Plane	A conductor layer, or portion of a conductor layer, used as a common reference point for circuit returns, shielding, or heat sinking.
Haloing	Mechanically induced fracturing delamination on or below the surface of the base material; it is usually exhibited by a light area around holes, or other machines areas, or both.
Hole Breakout	A condition in which a hole is not completely surrounded by the land.
Hole Density	The quantity of holes in a printed circuit board per unit area.
Hole Void	A void in the metallic deposit of a plated-through hole exposing the base material.
Image	That portion on artwork masters, working tools, silk screens, or photo masks that would be considered the photographic image. Also would include images created with photo-resists or silk-screening techniques. Generally, "one image" refers to a single circuit board image; thus there may be several images per flat.
Ink	Common term for screen resist.
Inner layer	Any layer that will be pressed on the inside of a multilayer board.
Inspection Overlay	A positive or negative transparency made from the production master and used as an inspection aid.
Insulation Resistance	The electrical resistance of the insulating material (determined under specified conditions) as measured between any pair of contacts or conductors.
Jumper Wire	An electrical connection formed by wire between two points on a printed board added after the intended conductive pattern is formed.
Kerf	A widening of the rout path as may be called out on the blueprint. Allows extra space for hardware to be attached to the board.

Term	Definition
Keying Slot	A slot in a printed circuit board that polarizes it, thereby permitting it to be plugged into its mating receptacle with pins properly aligned, but preventing it from being reversed or plugged into any other receptacle.
Laminate	A product made by bonding together two or more layers of material.
Laminate Thickness	Thickness of the base material, not including metal-clad, prior to any processing. Applies to single- or double-sided material.
Laminate Void	Absence of laminate material in an area that normally contains laminate material.
Laminating Presses, Multilayer	Equipment that applies both pressure and heat to laminate and pre-preg to make multilayer boards.
Lamination	The process of preparing a laminate.
Land	A portion of a conductive pattern usually, but not exclusively, used for the connection and/or attachment of components. Also called Pad.
Landless Hole	A plated-through hole without land(s). Also referred to as padless plated holes.
Layer-to-Layer Spacing	The thickness of dielectric material between adjacent layers or conductive circuitry in a multilayer printed circuit board.
Lay-Up	(1) The technique of registering and stacking layers of materials (laminate and pre-preg) for a multilayer board in preparation for the laminating cycle. (2) The laying out of repeat images on film to create multiple groups of circuit boards. (3) The laying out of multiple layers in preparation for multilayer lamination.
Major Defect	A defect that could result in a failure or significantly reduces the usability of the part for its intended purpose.
Mask	A material applied to enable selective etching, plating, or the application of solder to a printed circuit board.
Measling	Condition existing in the base laminate in the form of discrete white spots or "crosses" below the surface of the base laminate, reflecting a separation of fibers in the glass cloth at the weave intersection.
Metlab	Short for Metallurgical Laboratory. (1) Refers to the process(es) of inspecting internal board quality characteristics through the use of microsections. (2) Used interchangeably with *microsection.*
Microsectioning	The preparation of a specimen for the microscopic examination of the material to be examined, usually by cutting out a cross section, followed by encapsulation, polishing, etching, staining, and so on.
Microvia	A via used to make connection between two adjacent layers, typically less than 6 mils in diameter. May be formed by laser ablation, plasma etching, or photo processing.
Mil	One-thousandth of an inch (0.001″).
Minimum Annular Ring	The minimum metal width, at the narrowest point, between the circumference of the hole and the outer circumference of the land. This measurement is made to the drilled hole on internal layers of multilayer printed circuit boards and to the

Term	Definition
	edge of the plating on outside layers of multilayer boards and double-sided boards.
Minimum Electrical Spacing	The minimum allowable distance between adjacent conductors that is sufficient to prevent dielectric breakdown, corona, or both, between the semiconductors at any given voltage and altitude.
Minor Defect	A defect that is not likely to reduce the usability of the unit for its intended purpose. It may be a departure from established standards having no significant bearing on the effective use or operation of the unit.
Misregistration	The lack of dimensional conformity between successively produced features or patterns.
Multilayer Circuit Board	The general term for completely processed printed circuit configurations consisting of alternate layers of conductive patterns and insulating materials bonded together in more than two layers.
Nail Heading	The flared condition of copper on the inner conductor layers of a multilayer board caused by hole drilling.
Negative	An artwork master or production master in which the intended conductive pattern is transparent to light, and the areas to be free from conductive material are opaque.
Nonfunctional Land	A land on internal or external layers, not connected to the conductive pattern on its layer.
October Project	Industry consortium dedicated to waste reduction in circuit board manufacturing.
Outer Layer	A conductive layer that lies on the outside of a flat.
Outgassing	Deaeration or other gaseous emission from a printed circuit board when exposed to the soldering operation.
Overhang	Increase in printed circuit conductor width caused by plating build-up or by undercutting during etching.
Oxide	A chemical treatment to inner layers prior to lamination, for the purpose of increasing the roughness of clad copper to improve laminate bond strength.
Pad	The portion of the conductive pattern on printed circuits designated for the mounting or attachment of components. Also called Land.
Panel	The square or rectangular base material containing one or more circuit patterns that passes successively through the production sequence and from which printed circuit boards are extracted. *See Back Planes and Panels.*
Panel Plating	The electrolytic plating of the entire surface of a panel (including holes).
Part	Individual items that are not normally subject to disassembly without destruction.
Pattern	The configuration of conductive and nonconductive materials on a panel or printed board. Also the circuit configuration on related tools, drawings, and masters.

Term	Definition
Pattern Plating	Selective electrolytic plating of a conductive pattern.
Peel Strength	The force required to peel the conductor or foil from the base material.
Permittivity	Measure of the ability of a material to store electrical energy when exposed to an electrical field.
Photomask	A silver halide or diazo image on a transparent substrate that is used to either block or pass light.
Photo-resist	A light-sensitive material that is used to establish an image by exposure to light and chemical development.
Photo plotter	A high-accuracy (>0.002 inch) flatbed plotter with a programmable, photo image projector assembly. It is most often used to produce actual-size master patterns for printed circuit artwork directly on dimensionally stable, high-contrast photographic film.
Pilot Order	First production order going through process.
Pinhole	A minute hole through a layer or pattern.
Pitch	The nominal distance between the centers of adjacent features or traces on any layer of a printed circuit board. Also known as "center-to-center spacing."
Plated-Through Hole (PTH)	A hole in a circuit board that that been plated with metal (usually copper) on its sides to provide electrical connections between conductive patterns layers of a printed circuit board.
Plating	Chemical or electromechanical deposition of metal on a pattern.
Plating Resists	Material that, when deposited on conductive areas, prevents the plating of the covered areas. Resists are available both as screened-on materials and as dry-film photopolymer resists.
Plating Void	The absence of a plating metal from a specified plating area.
Plotting	The mechanical converting of x-y positional information into a visual pattern, such as artwork.
Polymide Resins	High-temperature thermoplastics used with glass to produce printed circuit laminates for multilayer and other circuit applications requiring high-temperature performance.
Polymerize	To unite chemically two or more monomers or polymers to form a molecule with a higher molecular weight.
Positional Limitation Tolerancing	Defines a zone within which the axis or center plane of a feature is permitted to vary from true (theoretically exact) position.
Pre-clean	Cleaning steps taken prior to an operation to ensure success of the operation.
Pre-preg	Sheet material consisting of the base material impregnated with a synthetic resin, such as epoxy or Polymide, partially cured to the B-stage (an intermediate stage). Short for pre-impregnated. *See also B-stage.*
Press-Fit Contact	An electrical contact that can be pressed into a hole in an insulator, printed board (with or without plated-through holes), or a metal plate.

Term	Definition
Printed Circuit	A conductive pattern of printed components and circuits attached to a common base.
Printed Circuit Board (PCB)	The general term for a printed or etched circuit board. It includes single, double, or multiple layer boards, both rigid and flexible.
Printed Wiring Board	A part manufactured from rigid base material upon which completely processed printed wiring has been formed.
Production Master	A 1 : 1 scale pattern that is used to produce one or more printed boards (rigid or flexible) within the accuracy specified on the master drawing. (1) Single-Image Product Master: A production master used in the process of making a single printed circuit board. (2) Multiple-Image Production Master: A production master used in the process of making two or more printed circuit boards simultaneously.
Reflowing	The melting of an electrodeposit followed by solidification. The surface has the appearance and physical characteristics of being hot-dipped.
Registration	The degree of conformity of the position of a pattern, or a portion thereof, with its intended position or with that of any other conductor layer of a board.
Residue	An undesirable substance remaining on a substrate after a process step.
Resin Smear	Resin transferred from the base material onto the surface or edge of the conductive pattern normally caused by drilling. Sometimes called epoxy smear.
Resin-Starved Area	A region in a printed circuit board that has an insufficient amount of resin to wet out the reinforcement completely evidenced by low gloss, dry spots, or exposed fibers.
Resist	Coating material used to mask or to protect selected areas of a pattern from the action of an etchant, solder, or plating. *Also see Dry-Film Resists, Plating Resists,* and *Solder Resists.*
Resistivity	The ability of a material to resist the passage of electrical current through it.
Reverse Image	The resist pattern on a printed circuit board enabling the exposure of conductive areas for subsequent plating.
Rework	Reprocessing that makes articles conform to specifications.
Robber	An exposed area generally attached to a rack used in electroplating, usually to provide a more uniform current density on plated parts. Thieves are intended to absorb the unevenly distributed current on parts, thereby assuring that the parts will receive a uniform electroplated coating.
Schematic Diagram	A drawing that shows, by means of graphic symbols, the electrical connections, components, and functions of an electronic circuit.
Screen	A cloth material (usually polyester or stainless steel for circuit boards) coated with a pattern that determines the flow and location of coatings forced through its openings.
Screen Printing	A process for transferring an image to a surface by forcing suitable media through a stencil screen with a squeegee. Also called silk screening.

Term	Definition
Selective Plate	A process for plating unique features with a different metal than the remaining features will have. Created by imaging, exposing, and plating selected area and then repeating the process for the remainder of the board.
Shadowing	A condition occurring during etchback in which the dielectric material, in contact with the foil, is incompletely removed although acceptable etchback may have been achieved elsewhere.
Silk Screening	*See Screen Printing.*
Single-Sided Board	Circuit board with conductors on only one side and no plated-through holes.
Solder Leveling	The process of dipping printed circuit boards into hot solder and leveling with hot air.
Solder Mask	A coating applied to a circuit board to prevent solder from flowing onto any areas where it's not desired or from bridging across closely spaced conductors.
Solder Masking Coating	Nonpreferred term for *resist.*
Solder Resists	Coatings that mask and insulate portions of a circuit pattern where solder is not desired.
Solderability Testing	The evaluation of a metal to determine its ability to be wetted by solder.
Squeegee	The tool used in silk screening that forces the resist or ink through the mesh.
Starvation, Resin	A deficiency of resin in base material that is apparent after lamination by the presence of weave texture, low gloss, or dry spots.
Step-and-Repeat	A method by which successive exposures of a single image are made to produce a multiple-image production master.
Strip	The chemical removal of developed photoresist or plated metal.
Substrate	*See Base Material.*
Subtractive Process	A process in printed circuit manufacturing where the product is built by the subtraction of an already existing metallic coating. The opposite of additive processing.
Test Coupon	A sample or test pattern normally made outside the actual board pattern that is used for testing to verify certain quality parameters without destroying the actual board.
Thief	*See Robber.*
Tooling Holes	Two specified holes on a printed circuit board used to position the board in order to mount components accurately.
Traveler	A "recipe" for the manufacture of a board. It "travels" with each order from start to finish. The traveler identifies each order and gives instructions for each step in the process. It also provides information for traceability and history.
Two-Sided Board	*See Double-Sided Board.*

Term	Definition
Underwriters Laboratory	Certifying agency for consumer electronics. *See also Underwriters Symbol.*
Underwriters Symbol	A logotype denoting that the product has been recognized by Underwriters Laboratory, Inc. (UL).
Via	A plated thru-hole that is used as an inner-layer connection but doesn't have component lead or other reinforcing material in it.
Void	The absence of substances in a localized area (e.g., air bubbles).
Wave Soldering	A process wherein assembled printed boards are brought in contact with a continuously flowing and circulating mass of solder.
Zero Defects Sampling	A statistical based attribute sampling plan (C = O) where a given sample of parts is inspected and any defects found are cause for rejection of the entire lot.

PCB Manufacturing Acronyms

Term	Definition
AEA	American Electronics Association; professional association in electronics.
ANSI	American National Standards Institute.
APICS	American Production and Inventory Control Society: professional certification.
BOM	Bill of material.
CAD	Computer-aided design: computer graphics systems to aid designers.
CAE	Computer-aided engineering.
CAM	Computer-aided manufacturing.
CCA	Capital commitment authorization.
CIM	Computer-integrated manufacturing: electronic integration of information systems.
CLCA	Closed loop correction action.
CLT	Compressed lead time: orders requested by the customer to be completed in shorter than standard time.

PCB industry acronyms courtesy Merix Corp. (*www.MERIX.com*)

Term	Definition
Cp, Cpk	Capability indices to measure process potential and process performance.
CRP	Capacity requirement planning.
DEQ	Department of Environmental Quality: state government regulatory agency.
DFM	Design for manufacturability.
DM	Direct metalization.
DRS	Design rule software.
ECB	Etched circuit board.
EPA	Environmental Protection Agency: federal government regulatory group.
EPC	Emergency process change: a form used to alert those who need to know when a manufacturing process must be changed immediately.
Er	Relative permitivity.
FY	Fiscal year.
HASL	Hot air solder leveling: a process that applies solder to enhance the component solderability to the circuitry (also known as HAL).
HR	Human resources department.
HRIS	Human resources information system.
IPC	Institute for Interconnecting and Packaging Electronic Circuits: a professional association.
IPC Design Council	Industry consortium of circuit board design professionals sponsored by the IPC.
ISO	International Standards Organization: standards writing and certification association.
ML	Multilayer: circuit board with more than one layer of material.
MSDS	Material safety data sheet.
NIST	National Institute of Standards and Technology: agency for calibration and technology.
NPI	New product introduction.
OCC	Organic coated copper (interchangeable with OSP).
OEM	Original equipment manufacturer.
OSHA	Occupational Safety and Health Agency: a federal government regulatory agency.
OSP	Organic solderability preservative.
P/N	Part number.
PCB	Printed circuit board.

Term	Definition
PWB	Printed wiring board.
PSI	Parts-specific information.
SME	Society of Manufacturing Engineers: professional association for manufacturing.
SMI	Surface Mount International: trade publication in electronics.
SMT	Surface mount technology.
SMTA	Surface Mount Technology Association: professional association in electronics.
SPC	Statistical process control: philosophy for continuous reduction of variation.
TQC	Total quality control.
TQM	Total quality management: philosophy of systematic quality improvement.
UL	Underwriters Laboratory.
UV	Ultraviolet.
WIP	Work in process: inventory of unfinished goods.

Electronic Terms

Term	Definition
A	Abbreviation for "ampere," a unit of electrical current.
Absorption	Loss or dissipation of energy as it travels through a medium. Example: radio waves lose some of their energy as they travel through the atmosphere.
AC	Abbreviation for "alternating current."
Acceptor Atoms	Trivalent atoms that accept free electrons from pentavalent atoms.
AC Coupling	Circuit that passes an AC signal while blocking a DC voltage.
AC/DC	Equipment that will operate on either an AC or DC power source.
AC Generator	Device used to transform mechanical energy into AC electrical power.
AC Load Line	A graph representing all possible combinations of AC output voltage and current for an amplifier.
AC Power Supply	Power supply that delivers an AC voltage.
Active Component	A component that changes the amplitude of a signal between input and output.

Electronic Terms courtesy Twisted Pair (*www.twysted-pair.com*)

Term	Definition
Active Filter	A filter that uses an amplifier in addition to reactive components to pass or reject selected frequencies.
Active Region	The region of BJT operation between saturation and cutoff used for linear amplification.
AC Voltage	A voltage in which the polarity alternates.
ADC	Abbreviation for "analog to digital converter."
Admittance	(symbol "Y") Measure of how easily AC will flow through a circuit. Admittance is the reciprocal of impedance and is measured in siemens.
AF	Abbreviation for "audio frequency."
AFC	Abbreviation for "automatic frequency control."
AGC	Abbreviation for "automatic gain control."
Alkaline Cell	A primary cell that delivers more current than a carbon-zinc cell. Also known as an "alkaline manganese cell."
Alligator Clip	Spring clip on the end of a test lead used to make a temporary connection.
Alpha	Ratio of collector current to emitter current in a bipolar junction transistor (BJT). Greek letter α is the symbol used.
Alternating Current	An electric current that rises to a maximum in one direction, falls back to zero, and then rises to a maximum in the opposite direction and then repeats.
Alternator	Name for an AC generator.
AM	Abbreviation for "amplitude modulation."
Ammeter	A meter used to measure current.
Ampere	Unit of electrical current.
Amplifier	A circuit that increases the voltage, current, or power of a signal.
Amplitude	Magnitude or size of a signal voltage or current.
Analog	Information represented as continously varying voltage or current rather than in discrete levels as opposed to digital data varying between two discrete levels.
Anode	The positive electrode or terminal of a device. The "P" material of a diode.
Antenna, Transmitting	A device that converts an electrical wave into an electromagnetic wave that radiates away from the antenna.
Antenna, Receiving	A device that converts a radiated electromagnetic wave into an electrical wave.
Apparent Power	Power attained in an AC circuit as a product of effective voltage and current that reach their peak at different times.
Arc	Discharge of electricity through a gas such as lightning discharging through the atmosphere.
Armature	The rotating or moving component of a magnetic circuit.

Term	Definition
Armstrong Oscillator	An oscillator that uses an isolation transformer to achieve positive feedback from output to input.
Astable Multivibrator	An oscillator that produces a square wave output from a DC voltage.
Atom	The smallest particle that an element can be broken down into and still maintain its unique identity.
Atomic Number	The number of positive charges or protons in the nucleus of an atom.
Attenuate	To reduce the amplitude of an action or signal. The opposite of amplification.
Audio	Relating to frequencies that can be heard by the human ear. Approximately 20 Hz to 20 kHz.
Autotransformer	A single winding transformer where the output is taken from taps on the winding.
Average Value	A value of voltage or current where the area of the wave above the value equals the area of the wave below the value.
AVC	Abbreviation for "automatic volume control."
Avionics	Aviation electronics.
AWG	Abbreviation for "american wire gauge." A gauge that assigns a number value to the diameter of a wire.
Balanced Bridge	Condition that occurs when a bridge circuit is adjusted to produce a zero output.
Band-pass Filter	A tuned circuit designed to pass a band of frequencies between a lower cut-off frequency (f_1) and a higher cut-off frequency (f_2). Frequencies above and below the pass band are heavily attenuated.
Band-stop Filter	A tuned circuit designed to stop frequencies between a lower cut-off frequency (f_1) and a higher cut-off frequency (f_2) of the amplifier while passing all other frequencies.
Bandwidth	Width of the band of frequencies between the half power points.
Barrier Potential	The natural difference of potential that exists across a forward biased PN junction.
Base	The region that lies betwen the emitter and collector of a bipolar junction transistor (BJT).
Base Biasing	A method of biasing a BJT in which the bias voltage is supplied to the base by means of a resistor.
Battery	A DC voltage source containing two or more cells that convert chemical energy to electrical energy.
Baud	A unit of signaling speed equal to the number of signal events per second. Not necessarily the same as bits per second.
Beta	β; The ratio of collector current to base current in a bipolar junction transistor (BJT).

Term	Definition
Bias	A DC voltage applied to a device to control its operation.
Binary	A number system having only two symbols, 0 and 1. A base 2 number system.
Bipolar Junction Transistor (BJT)	A three-terminal device in which emitter to collector current is controlled by base current.
Bistable Multivibrator	A multivibrator with two stable states. An external signal is required to change the output from one state to the other. Also called a latch.
Bleeder Current	A current drawn continously from a souce. Bleeder current is used to stabilize the output voltage of a source.
Bode Plot	A graph of gain versus frequency.
Branch Current	The portion of total current flowing in one path of a parallel circuit.
Breakdown Voltage	Voltage at which the breakdown of a dielectric or insulator occurs.
Breakover Voltage	Minimum voltage required to cause a diac to break down and conduct.
Bridge Rectifier	A circuit using four diodes to provide full wave rectification. Converts an AC voltage to a pulsating DC voltage.
Buffer	An amplifier used to isolate a load from a source.
Bulk Resistance	The natural resistance of a P type or N type semiconductor material.
Butterworth Filter	A type of active filter characterized by a constant gain (flat response) across the midband of the circuit and a 20-dB-per-decade roll-off rate for each pole contained in the circuit.
BW	Abbreviation for bandwidth.
Bypass Capacitor	A capacitor used to provide an AC ground at some point in a circuit.
Byte	Group of eight binary digits or bits.
Cable	Group of two or more insulated wires.
CAD	Abbreviation for "computer aided design."
Calibration	To adjust the correct value of a reading by comparison to a standard.
Capacitance	The ability of a capacitor to store an electrical charge. The basic unit of capacitance is the Farad.
Capacitive Reactance	The opposition to current flow provided by a capacitor. Capacitive reactance is measured in ohms and varies inversly with frequency.
Capacitor	An electronic component having capacitive reactance.
Capacitor Microphone	Microphone whose operation depends on variations in capacitance caused by varying air pressure on the movable plate of a capacitor.
Carbon-film Resistor	Device made by depositing a thin carbon film on a ceramic form.
Carbon Microphone	Microphone whose operation depends on pressure variation in carbon granules causing a change in resistance.

Term	Definition
Carbon Resistor	Resistor of fixed value made by mixing carbon granules with a binder which is moulded and then baked.
Cascaded Amplifier	An amplifier with two or more stages arranged in a series configuration.
Cascode Amplifier	A high-frequency amplifier made up of a common-source amplifier with a common-gate amplifier in its drain network.
Cathode	The negative terminal electrode of a device. The N material in a junction diode.
Cathode Ray Tube (CRT)	Vacuum tube used to display data in a visual form. Picture tube of a television or computer terminal.
Cell	Single unit used to convert chemical energy into a DC electrical voltage.
Center Frequency	Frequency to which an amplifier is tuned. The frequency halfway between the cut-off frequencies of a tuned circuit.
Center Tap	Midway connection between the two ends of a winding.
Center Tapped Rectifier	Circuit that make use of a center tapped transformer and two diodes to provide full wave rectification.
Center Tapped Transformer	A transformer with a connection at the electrical center of a winding.
Ceramic Capacitor	Capacitor in which the dielectric is ceramic.
Charge	Quantity of electrical energy.
Charge Current	Current that flows to charge a capacitor or battery when voltage is applied.
Chassis	Metal box or frame into which components are mounted.
Chassis Ground	Connection to a chassis.
Chebyshev Filter	A type of active filter characterized by high roll-off rates (40 dB per decade per pole) and midband gain that is not constant.
Choke	Inductor used to oppose the flow of alternating current.
Circuit	Interconnection of components to provide an electrical path between two or more components.
Circuit Breaker	A protective device used to open a circuit when current exceeds a maximum value. In effect, a reusable fuse.
Clamper	A diode circuit used to change the DC level of a waveform without distorting the waveform.
Clapp Oscillator	A variation of the Colpitts oscillator. An added capacitor is used to eliminate the effects of stray capacitance on the operation of the basic Colpitts oscillator.
Class A Amplifier	A linear amplifier biased so the active device conducts through 360 degrees of the input waveform.
Class B Amplifier	An amplifier with two active devices. The active components are biased so that each conducts for approximately 180 degrees of the input waveform cycle.

Term	Definition
Class C Amplifier	An amplifier in which the active device conducts for less than 180 degrees of the input waveform cycle.
Clipper	A diode circuit used to eliminate part of a waveform.
Clipping	Distortion caused by overdriving an amplifier.
Clock	A square waveform used for synchronizing and timing of several circuits.
Cosed Circuit	Circuit having a complete path for current flow.
Closed-Loop Gain	Gain of an amplifier when a feedback path is present.
Coaxial Cable	Transmission line in which the signal carrying conductor is covered by a dielectric and another conductor.
Coefficient of Coupling	The degree of coupling between two circuits.
Coercive Force (H)	Magnetizing force needed to reduce residual magnetism in a material to zero.
Collector	The semiconductor region in a bipolar junction transistor through which a flow of charge carriers leaves the base region.
Collector Characteristic Curve	A graph of collector voltage over collector current for a given base current.
Color Code	Set of colors used to indicate value of a component.
Colpitts Oscillator	An oscillator with a pair of tapped capacitors in the feedback network.
Common-Anode Display	A multisegment light emitting diode (LED) with a single positive voltage input connection. Separate cathode connections are provided for each individual segment.
Common Cathode Display	A multisegment light emitting diode (LED) with a single negative voltage input connection. Separate anode connections are provided for each individual segment.
Common Base Amplifier	A BJT circuit in which the base connection is common to both input and output.
Common Collector Amplifier	A BJT circuit in which the collector connection is common to both input and output.
Common Drain Amplifier	An FET circuit in which the drain connection is common to both input and output.
Common Emitter Amplifier	A BJT circuit in which the emitter connection is common to both input and output.
Common Gate Amplifier	An FET circuit in which the gate connection is common to both input and output.
Common Source Amplifier	An FET circuit in which the source connection is common to both input and output.

Term	Definition
Common-Mode Rejection Ratio (CMRR)	The ratio of op-amp differential gain to common-mode gain. A measure of an op-amp's ability to reject common-mode signals such as noise.
Common-Mode Signals	Signals that appear simultaneously at two inputs of an operational amplifier (op-amp). Common mode signals are always equal in amplitude and phase.
Comparator	An op-amp circuit that compares two inputs and provides a DC output indicating the polarity relationship between the inputs.
Complementary Symmetry Amplifier	A class B amplifier using matched complementry transistors. Does not require a phase inverter for push-pull output.
Complementry Transistors	Two transistors, one *NPN* and one *PNP,* having near identical charastics. *N*-channel and *P*-channel FETs can also be complementry.
Complex Numbers	Numbers composed of a real number part and an imaginary number part.
Compliance	The maximum possible peak-to-peak output of an amplifier.
Constant Current Circuit	Circuit used to maintain constant current to a load having resistance that changes.
Contact	Current carrying part of a switch, relay or connector.
Continuity	Occurs when a complete path for current exists.
Conventional Current Flow	Concept of current produced by the movement of positive charges towards the negative terminal of a source.
Copper Loss	Power lost in transformers, generators, connecting wires and other parts of a circuit due to current flow through the resistance of copper conductors.
Core	Magnetic material within a coil used to concentrate the magnetic field.
Coulomb	Unit of electric charge. A negative coulomb charge consists of 6.24×10^{18} electrons.
Counter Electromotive Force (Counter EMF)	Voltage induced into an inductor due to an alternating or pulsating current. Counter emf is always in polarity opposite to that of the applied voltage. Opposing a change of current.
Coupling	To electronically connect two circuits so that signal will pass from one to the other.
Covalent Bond	The way some atoms complete their valence shells by sharing valence electrons with neighboring atoms.
Crossover Distortion	Distortion caused by both devices in a class B amplifier being cut off at the same time.
Crowbar	Circuit used to protect the output of a souce from a short circuited load. Load current is limited to a value the source can deliver without damage.
CRT	Abbreviation for "cathode ray tube."
Crystal	Natural or synthetic piezoelectric or semiconductor material with atoms arranged with some degree of geometric regularity.

Term	Definition
Crystal-Controlled Oscillator	Oscillator that uses a quartz crystal in its feedback path to maintain a stable output frequency.
Current	Measured in amperes, it is the flow of electrons through a conductor. Also know as electron flow.
Current Amplifier	Amplifier to increase signal current.
Current Divider	Parallel network designed to divide the total current of a circuit.
Current Feedback	Feedback configuration where a portion of the output current is fed back to the amplifier input.
Current-Limiting Resistor	Resistor in the path of current flow to control the amount of current drawn by a device.
Current Mirror	Term used to describe the fact that DC current through the base circuit of a class B amplifier is approximately equal to the DC collector current.
Cutoff	Condition when an active device is biased such that output current is near zero or beyond zero.
Cutoff Frequency	Frequency at which the power gain of an amplifier falls below 50% of maximum.
Cycle	When a repeating wave rises from zero to a positive maximum then back to zero and on to a negative maximum and back to zero it is said to have completed one cycle.
DAC	Abbreviation for "digital to analog converter."
Damping	Reduction in magnitude of oscillation due to energy being dissipated as heat.
Darlington Pair	An amplifier consisting of two bipolar junction transistors with their collectors connected together and the emitter of one connected to the base of the other. Circuit has an extremely high current gain and input impedance.
DC	Abbreviation for "direct current."
DC Load Line	A graph representing all possible combinations of voltage and current for a given load resistor in an amplifier.
DC Offset	The change in input voltage required to produce a zero output voltage when no signal is applied to an amplifier.
DC Power Supply	Any source of DC power for electrical equipment.
Dead Short	Short circuit having zero resistance.
Decade	A frequency factor of ten.
Decibel (dB)	A logarithmic representation of gain or loss.
Degenerative Feedback	Also called negative feedback. A portion of the output of an amplifier is inverted and connected back to the input. This controls the gain of the amplifier and reduces distortion and noise.
Delay Time	The time for collector current to reach 10% of its maximum value in a BJT switching circuit.

Term	Definition
Depletion Region	The area surrounding a *pn* junction that is depleted of carriers.
Depletion Mode	In an FET, an operating mode where reverse gate-source voltage is used to deplete the channel of free carriers. This reduces the size of the channel and increases its resistance.
Depletion-Mode MOSFET	A MOSFET designed to operate in either depletion mode or enhancement mode.
Device	A component or part.
Diac	A two terminal bidirectional thyristor. Has a symmetrical switching mode.
Dielectric	Insulating material between two plates where an electrostatic field exists.
Dielectric Constant	Peoperty of a material that determines how much electrostatic energy can be stored per unit volume when unit voltage is applied.
Dielectric Strength	The maximum voltage an insulating material can withstand without breaking down.
Differential Amplifier	An amplifier in which the output is in proportion to the differences between voltages applied to its two inputs.
Differentiator	A circuit in which the output voltage is in proportion to the rate of change of the input voltage. A high-pass RC circuit.
Diffusion	Tendency of conduction band electrons to wander across a pn junction to combine with valence band holes.
Digital	Relating to devices or circuits that have outputs of only two discrete levels. Examples: 0 or 1, high or low, on or off, true or false.
Diode	A two-terminal device that conducts in only one direction.
DIP	Abbreviation for "dual in-line package."
Direct Coupling	Where the output of an amplifier is connected directly to the input of another amplifier or to a load. Also known as DC coupling because DC signals are not blocked.
Direct Current	Current that flows in only one direction.
Discharge	Release of energy stored in either a battery or a capacitor.
Discrete Component	Package containing only a single component as opposed to an integrated circuit containg many components in a single package.
Dissipation	Release of electrical energy in the form of heat.
Distortion	An undesired change in a waveform or signal.
Distributed Capacitance	Any capacitance other than that within a capacitor. For example, the capacitance between adjacent turns of wire in a coil.
Distributed Inductance	Any inductance other than that within an inductor. Example: inductance in any conductor.

Term	Definition
Domain	A moveable magnetized area in a magnetized material. Also known as magnetic domain.
Donor Atoms	Pentavalent atoms that give up electrons to the conduction band in an N-type semiconductor material.
Doping	The process of adding impurity atoms to intrinsic (pure) silicon or germanium to improve the conductivity of the semiconductor material.
Dot Convention	Standard used with transformer symbols to indicate whether the secondary voltage is in phase or out of phase with the primary voltage.
Drift	A problem that can develop in tuned amplifiers when the frequency of the tuned circuit changes due to temperature or component aging.
Dropping Resistor	Resistor whose value has been chosen to drop or develop a given voltage.
Dry Cell	DC voltage generating chemical cell using a non liquid (paste) electrolyte.
Dual In-Line Package	Integrated circuit package having two rows of connecting pins.
Dual Trace Oscilloscope	Oscilloscope that can simultaneously display two signals.
Dynamic	Relating to conditions that are changing or in motion.
E-core	Laminated form in the shape of the letter "E," onto which inductors and transformers are wound.
Eddy Currents	Currents induced into a conducting core due to the changing magnetic field. Eddy currents produce heat which is a loss of power and lowers the efficiency of an inductor.
Efficiency	The amount of power delivered to the load of an amplifier as a percentage of the power required from the power supply.
Electric Charge	Electric energy stored on the surface of a material. Also known as a static charge.
Electric Field	A field or force that exists in the space between two different potentials or voltages. Also known as an electrostatic field.
Electricity	Science states that certain particles possess a force field or charge. The charge possessed by an electron is negative while the charge possessed by a proton is positive. Electricity can be divided into two groups, static and dynamic. Static electricty deals with charges at rest and dynamic electricity deals with charges in motion.
Electric Polarization	A displacement of bound charges in a dielectric when placed in an electric field.
Electroacoustic Transducer	Device that produces an energy transfer from electric to acoustic (sound) or from acoustic to electric. Examples include a microphone, earphones, and loudspeakers.
Electroluminescence	Conversion of electrical energy into light energy.
Electrolyte	Electrically conducting liquid (wet) or paste (dry).

Term	Definition
Electrolytic Capacitor	A capacitor having an electrolyte between the two plates. A thin layer of oxide is deposited on only the positive plate. The oxide acts as the dielectric for the capacitor. Electrolytic capacitors are polarized and so must be connected in correct polarity to prevent breakdown.
Electromagnet	A coil of wire usually wound on a soft iron or steel core. When current is passed through the coil, a magnetic field is generated. The core provides an easy path for the magnetic lines of force. This concentrates the field in the core.
Electromagnetic Communication	Use of an electromagnetic wave to pass information between two points. Also called wireless communication.
Electromagnetic Induction	Voltage produced in a coil due to relative motion between the coil and magnetic lines of force.
Electromagnetic Spectrum	List or diagram showing the range of electromagnetic radiation.
Electromagnetic Wave	Wave that consists of both electric and magnetic variation.
Electromagnetism	Relates to the magnetic field generated around a conductor when current is passed through it.
Electromechanical Transducer	Device that transforms electrical energy into mechanical energy (electric motor) or mechanical energy into electrical energy (generator).
Electromotive Force (EMF)	Force that causes the motion of electrons due to potential difference between two points. (voltage)
Electron	Smallest subatomic particle of negative charge that orbits the nucleus of an atom.
Electron Flow	Electrical current produced by the movement of free electrons toward a positive terminal.
Electrostatic	Related to static electric charge.
Electrostatic Field	Force field produced by static electrical charges.
Emitter	The semiconductor region from which charge carriers are injected into the base of a bipolar junction transistor.
Emitter Feedback	Coupling from the emitter output to the base input of a bipolar junction transistor.
Emitter Follower	A common collector amplifier. Has a high current gain, high input impedance, and low output impedance.
Energized	Being electrically connected to a voltage source so the device is activated.
Energy	Capacity to do work.
Engineering Notation	A floating point system in which numbers are expressed as products consisting of a number greater than one multiplied by an appropriate power of 10 that is some multiple of 3.
Enhancement-Mode MOSFET	A field effect transistor in which there are no charge carriers in the channel when the gate source voltage is zero.

Term	Definition
Equivalent Resistance	Total resistance of all the individual resistances in a circuit.
Fall Time	Time it takes the falling edge of a pulse to go from 90% of peak voltage to 10% of peak voltage.
Farad	The basic unit of capacitance.
Feedback	A portion of the output signal of an amplifier that is connected back to the input of the same amplifier.
Feedback Amplifier	An amplifier with an external signal path from its output back to its input.
Ferrite	A powdered, compressed, and sintered magnetic material having high resistivity. The high resistance makes eddy current losses low at high frequencies.
Ferrite Bead	Ferrite composition in the form of a bead. Running a wire through the bead increases the inductance of the wire.
Ferrite-Core Inductor	An inductor wound on a ferrite core.
Ferrites	Compound composed of iron oxide, a metallic oxide, and ceramic. The metal oxides include zinc, nickel, cobalt, or iron.
Ferrous	Composed of and or containing iron. A ferrous metal exhibits magnetic characteristics as opposed to nonferrous material.
Fiber Optics	Laser's light output carries information that is conveyed between two points by thin glass optical fibers.
Field Effect Transistor (FET)	A voltage controlled transistor in which the source to drain conduction is controlled by gate to source voltage.
Filament	Thin thread of carbon or tungsten which produces heat or light with the passage of current.
Filter	Network consisting of capacitors, resistors and/or inductors used to pass certain frequencies and block others.
Flip Flop	A bistable multivibrator. A circuit which has two output states and is switched from one to the other by means of an external signal (trigger).
Floating Ground	Conmmon connection in a circuit that provides a return path for current but is not connected to an earth ground.
Flow Soldering	Flow or wave soldering technique in large scale electronic assembly to solder all the connections on a printed circuit board by moving the board over a wave of molten solder.
Flux	Material used to remove oxide films from the surface of metals in preparation for soldering.
Flux	In magnetism, the magnetic field consisting of lines of force.
Flux Density	The concentration of magnetic lines of force. Determines strength of the magnetic field.
Flywheel Effect	Sustaining effect of oscillation in an LC circuit.

Term	Definition
Forward Bias	A PN junction bias which allows current to flow through the junction. Forward bias decreases the resistance of the depletion layer.
Free Electrons	Electrons that are not in any orbit around a nucleus.
Free Runing Multivibrator	A multivibrator that produces a continuous output waveform without any signal input. A square wave generator used to produce a clock signal.
Frequency	Rate of recurrence of a periodic wave. Measured in Hertz (cycles per second).
Frequency-Division Multiplex (FDM)	Transmission of two or more signals over a common path by using a different frequency band for each signal.
Frequency-Domain Analysis	A method of representing a waveform by plotting its amplitude against frequency.
Frequency Meter	Meter used to measure frequency of periodic waves.
Frequency Multiplier	A harmonic conversion circuit in which the frequency of the output signal is an exact multiple of the input frequency.
Frequency Response	Indication of how well a circuit responds to different frequencies applied to it.
Frequency Response Curve	A graph of amplitude over frequency indicating a circuit response to different frequencies.
Full-Scale Deflection	(FDS) Deflection of a meter's pointer to the farthest position on the scale.
Full-Wave Rectifier	Rectifier that makes use of the full AC wave in both the positive and negative half cycles.
Function Generator	Signal generator that can produce sine, square, triangle, and sawtooth output waveforms.
Fundamental Frequency	Lowest frequency in a complex waveform.
Fuse	A protective device in the current path that melts or breaks when current exceeds a predetermined maximum value.
Gain	Increase in voltage, current and/or power. Gain is expressed as a ratio of amplifier output value to the corresponding amplifier input value.
Gain Bandwidth Product	A device parameter that indicates the maximum possible product of gain and bandwidth. The gain bandwidth product of a device is equal to the *unity gain frequency* (f_{unity}) of the device.
Gamma Rays	High-frequency electromagnetic radiation from radio active particles.
Ganged	Mechanical coupling of two or more capacitors, switches, potentiometers, or any other adjustable components so that adjusting one control will operate all.
Gas	Any aeriform or completely elastic fluid which is not a solid or a liquid. Gasses are produced by heating a liquid beyond its boiling point.
Geiger Counter	Device used to detect nuclear particles.
Generator	Device used to convert mechanical energy to electrical energy.

Term	Definition
Giga	Metric prefix for 1 billion (10^9).
Ground	An intentional or accidental conducting path between an electrical system or circuit and the earth or some conducting body acting in place of the earth. A ground is often used as the common wiring point or reference in a circuit.
Gunn Diode	A semiconductor diode that utilizes the Gunn effect to produce microwave frequency oscillation or to amplify a microwave frequency signal.
H-Parameters	(hybrid parameters) Transistor specifications that describe the component operating limits under specific circumstances.
Half Power Point	A frequency at which the power is 50% of maximum. This corresponds to 70.7% of maximum current or voltage.
Half Wave Rectifier	A diode rectifier that converts AC to pulsating DC by eliminating either the negative or the positive alternation of each input AC cycle.
Harmonic	Sine wave that is smaller in amplitude and some multiple of a fundamental frequency. *Example:* 880 Hz is the second harmonic of 440 Hz, 880 Hz is the third harmonic of 220 Hz.
Hartley Oscillator	An oscillator that uses a tapped inductor in the feedback network.
Henry	The basic unit of inductance.
Hertz (Hz)	Unit of frequency. One hertz is equal to one cycle per second.
High Fidelity (Hi Fi)	Sound reproduction equipment that reproduces sound as near to the original sound as possible.
High-Pass Filter	A tuned circuit designed to pass all frequencies above a desnigated cut-off frequency. Frequencies below the cut-off frequency are rejected or attenuated.
High Tension	Lethal voltage in the kilovolt range and above.
Hole	A gap left in the covalent bond when a valence electron gains sufficient energy to jump to the conduction band.
Hologram	Three-dimensional picture created with a laser.
Holography	The science dealing with three-dimensional optical recording.
Horizontally Polarized Wave	Electromagnetic wave that has the electric field in the horizontal plane.
Hybrid Circuit	Circuit that combines two technologies (passive and active or discrete and integrated components) onto one microelectronic circuit. Passive components are usually made by thin film techniques, while active components are made with semiconductor techniques.
Hysteresis	Amount that the magnetization of a meterial lags the magnetizing force due to molecular friction. In Schmitt trigger circuits, the difference between the upper and lower trigger points.
IC	Abbreviation for "integrated circuit."

Term	Definition
IC Voltage Regulator	Three-terminal device used to hold the output voltage of a power supply constant over a wide range of load variations.
IGFET	Insulated gate field effect transistor. Another name for a "MOSFET."
Impedance	(Z) Measured in ohms it is the total opposition to the flow of current offered by a circuit. Impedance consists of the vector sum of resistance and reactance.
Impedance Coupling	Coupling of two signal amplifier circuits through the use of an impedance such as a inductor.
Impedance Matching	Matching the output impedance of a source to the input impedance of a load to attain maximum power transfer.
Incandescence	State of a material when heated to the point where it emits light (red hot or white hot).
Induced Voltage	Voltage generated in a conductor when subjected to a moving magnetic field.
Inductance	Property of a circuit to oppose a change in current. The moving magnetic field produced by a change in current causes an induced voltage to oppose the original change.
Inductive Circuit	Circuit having greater inductive reactance than capacitive reactance.
Inductive Reactance	Opposition to the flow of AC current produced by an inductor. Measured in Ohms and varies in direct proportion to frequency.
Inductor	Length of conductor used to introduce inductance into a circuit. The conductor is usually wound into a coil to concentrate the magnetic lines of force and maximize the inductance. While any conductor has inductance, in common usage the term inductor usually refers to a coil.
Infrared	Electromagnetic heat radiation whose frequencies are above the microwave frequency band and below red in the visible band.
Inhibit	To stop an action or block data from passing.
In Phase	When two or more waves of the same frequency have their positive and negative peaks occuring at the same time.
Input Impedance	Opposition to the flow of signal current at the input of a circuit or load.
Insulated	When a nonconducting material is used to isolate conducting materials from one another.
Insulating Material	Material that will prevent the flow of current due to its chemical composition.
Insulation Resistance	Resistance of insulating material. The greater the insulation resistance, the better the insulation.
Integrated	When two or more components are combined into a circuit and then incorporated into a single package.
Integrator	A device that approximates and whose output is proportional to an integral of the input signal. A low pass filter.
Intermediate Frequency Amplifier	In a superheterodyne radio it amplifies a fixed frequency lower than the received radio frequency and higher than the audio frequency.

Term	Definition
Intermittent	A fault occuring at random intervals of time. Intermittent problems are often difficult to locate because of the random nature. They often don't occur when the technician is present.
Internal Resistance	Every source has some resistance in series with the output current. When current is drawn from the source some power is lost due to the voltage drop across the internal resistance. Usually called output impedance or output resistance.
Intrinsic Material	A semiconductor material with electrical properties essentially characteristic of ideal pure crystal. Essentially silicon or germanium crystal with no measurable impurities.
Intrinsic Stand-Off Ratio	A unijunction transistor (UJT) rating used to determine the firing potential of the device.
Inverting Amplifier	An amplifier that has a 180° phase shift from input to output.
Inverting Input	In an operational amplifier (op-amp) the input that is marked with a minus sign. A signal applied at the inverting input will be given 180° phase shift between input and output.
Ion	An atom with fewer electrons in orbit than the number of protons in the nucleus is a positive ion. An atom with a greater number of electrons in orbit than the number of protons in the nucleus is a negative ion.
Ionized	Atoms become ionized when they gain or lose a valence electron.
j	A prefix used to indicate an imaginary number (operator j).
Jack	Socket or connector into which a plug may be inserted.
JFET	Abbreviation for "junction field effect transistor."
Joule	The unit of work and energy.
Junction	Contact or connection between two or more wires or cables. The area where the *P*-type material and *N*-type material meet in a semiconductor.
Junction Diode	A semiconductor diode in which the rectifying characteristics occur at a junction between the *n*-type and *p*-type semiconductor materials.
Kilo	Metric prefix for 1000 (10^3).
Kilovolt-Ampere	1000 volts at 1 ampere.
Kilowatt-Hour	1000 watts for 1 hour.
Kilowatt-Hour Meter	A meter used by electric utility companies to measure the amount of electric power used by a customer.
Kinetic Energy	Energy associated with motion.
Kirchoff's Current Law	The sum of the currents flowing into a point in a circuit is equal to the sum of the currents flowing out of that same point.
Kirchoff's Voltage Law	The algebraic sum of the voltage drops in a closed path circuit is equal to the algebraic sum of the source voltages applied.

Term	Definition
Knee Voltage	The voltage at which a curve joins two relatively straight portions of a characteristic curve. For a *PN* junction diode, the point in the forward operating region of the characteristic curve where conduction starts to increase rapidly. For a zener diode, the term is often used in reference to the zener voltage rating.
Lag	Difference in time between two waveforms of the same frequency expressed in degrees. Example: One waveform lags another waveform by a certain number of degrees.
Laminated Core	Core made up of sheets of magnetic material insulated from one another by an oxide or varnish.
Lamp	Device that produces light.
Laser	Device that produces a very narrow intense beam of light. The name is an acronym for "light amplification by stimulated emission of radiation."
Lead	The angle by which one alternating signal leads another in time. Opposite of lag. Also a wire that connects two points in a circuit.
Lead-Acid Cell	Cell made up of lead plates immersed in a sulphuric acid electrolyte. An automobile battery usually consists of six lead-acid cells.
Leakage	Small undesireable flow of current through an insulator or dielectric.
LED	Abbreviation for "light emitting diode."
Left-Hand-Rule	If fingers of the left hand are placed around a wire so that the thumb points in the direction of electron flow, the fingers will be pointing in the direction of the magnetic field being produced by the conductor.
Lenz's Law	The current induced in a circuit due to a change in the magnetic field is so directed as to oppose the flux, or to exert a mechanical force to oppose the motion.
Level Detector	An op-amp circuit that compares two inputs and provides a DC output indicating the polarity relationship between the inputs. A comparitor.
Lie Detector	Piece of electronic equipment also called a polygraph used to determine whether a person is telling the truth by looking for dramatic changes in blood pressure, body temperature, breathing rate, heart rate and skin moisture in response to questions.
Light	Electromagnetic radiation in a band of frequencies that can be received by the human eye.
Lifetime	The time from the creation of an electron hole pair until recombination occurs.
Light-Emitting Diode	A semiconductor diode that converts electric energy into electromagnetic radiation at a visible and near infrared frequencies when its *pn* junction is forward biased.
Limiter	Circuit or device that prevents some portion of its input from reaching the output. A clipper.
Linear	Relationship between input and output in which the output varies in direct proportion to the input.
Linear Scale	A scale in which the divisions are uniformly spaced.

Term	Definition
Line Regulation	The ability of a voltage regulator to maintain a constant voltage when the regulator input voltage varies.
Live	Term used to describe a circuit or piece of equipment that is on and has current flow within it.
Load	A source drives a load. Whatever component or piece of equipment is connected to a source and draws current from a source is a load on that source.
Load Current	Current drawn from a source by a load.
Load Impedance	Vector sum of reactance and resistance in a load.
Loading Effect	Large load impedance will draw a small load current and so loading of the source is small (light load). A small load impedance will draw a large load current from the source (heavy load).
Load Regulation	The ability of a voltage regulator to maintain a constant output voltage under varying load currents.
Load Resistance	Resistance of a load.
Logic	Science of dealing with the principle and applications of gates, relays, and switches.
Loss	Term used to describe a decrease in power.
Low-Pass Filter	A tuned circuit designed to pass all frequencies below a designated cut-off frequency.
Magnet	Body that can be used to attract or repel magnetic materials.
Magnetic Circuit Breaker	Circuit breaker that is tripped or activated by use of an electromagnet.
Magnetic Coil	Spiral of a conductor that is called an electromagnet.
Magnetic Core	Material that exists in the center of the magnetic coil either to physically support the windings (nonmagnetic material) or to concentrate the magnetic flux (magnetic material).
Magnetic Field	Magnetic lines of force travelling from the north pole to the south pole of a magnet.
Magnetic Flux	The magnetic lines of force produced by a magnet.
Magnetic Leakage	The passage of magnetic flux outside the path along which it can do useful work.
Magnetic Poles	Points of a magnet from which magnetic lines of force leave (north pole) and arrive (south pole).
Magnetism	Property of some materials to attract or repel others.
Magnetizing Force	Also called magnetic field strength. It is the magnetomotive force per unit length at any given point in a magnetic circuit.
Magnetomotive Force	Force that produces a magnetic field.

Term	Definition
Majority Carriers	The conduction band electrons in an *n*-type material and the valence band holes in a *p*-type material. Produced by pentavalent impurities in *n*-type material and trivalent impurities in *p*-type material.
Matched Impedance	Condition that occurs when the output impedance of a souce is equal to the input impedance of a load.
Matching	Connection of two components or circuits so that maximum power is transferred between the two.
Maximum Power Transfer	A theorem that states that maximum power will be transferred from source to load when input impedance of the load equals the output impedance of the source.
Maxwell	Unit of magnetic flux. One maxwell equals one magnetic line of force.
Mercury Cell	Primary cell using a mercuric oxide cathode, a zinc anode, and a potassium hydroxide electrolyte.
Metal Film Resistor	A resistor in which a film of metal oxide or alloy is deposited on an insulating substrate.
Metal Oxide Field Effect Transistor (MOSFET)	A field effect transistor in which the insulating layer betwen the gate electrode and the channel is a metal oxide layer.
Metal Oxide Resistor	A metal film resistor in which an oxide of metal (such as tin) is deposited as a film onto the substrate.
Meter	Any electrical or electronic measuring device. In the metric system, it is the unit of length equal to 39.37 inches.
Meter FSD Current	Value of meter current needed to cause the needle to deflect to its maximum position (full-scale deflection).
Meter Resistance	DC resistance of the meter's armature coil.
Mica Capacitor	Capacitor using mica as the dielectric.
Microphone	Electroacoustic transducer that converts sound energy into elecric energy.
Microwave	Band of very short wavelength radio waves within the UHF, SHF, and EHF bands.
Midband Gain	Gain of an amplifier operating within its bandwidth.
Midpoint Bias	An amplifier biased at the center of its DC load line.
Mil	One thousandth of an inch (0.001 in.)
Miller's Theorem	A theorem that allows you to represent a feedback capacitor as equivalent input and output shunt capacitors.
Minority Carriers	The conduction band holes in *n*-type material and valence band electrons in *p*-type material. Most minority carriers are produced by temperature rather than by doping with impurities.

Term	Definition
Mismatch	Term used to describe a difference between the output impedance of a source and the input impedance of a load. A mismatch prevents the maximum transfer of power from source to load.
Modulation	Process by which an information signal (audio for example) is used to modify some characteristic of a higher frequency wave known as a carrier (radio, for example).
Monostable Multivibrator	A nultivibrator with one stable output state. When triggered, the circuit output will switch to the unstable state for a predetermined period of time and then return to the stable state. A timer.
Molecule	Smallest particle of a compound that still retains its characteristics.
MOSFET	Abbreviation for "metal oxide field effect transistor" (also known as an "insulated gate field effect transistor). A field effect transistor in which the insulating layer betwen the gate electrode and the channel is a metal oxide layer.
Moving Coil Microphone	Microphone that uses a moving coil within a fixed magnetic field. Dynamic microphone.
Moving Coil Pick-Up	Dynamic phonograph pick-up in which the stylus causes a coil to move within a fixed magnetic field.
Moving Coil Loudspeaker	Loudspeaker that uses a moving "voice coil" placed within a fixed magnetic field. Audio frequency current in the voice coil causes movement which is mechanically transferred to the speaker cone. Also known as a dynamic loudspeaker.
Multimeter	Electronic test equipment that can perform multiple tasks. Typically one capable of measuring voltage, current, and resistance. More sophisticated modern digital multimeters also measure capacitance, inductance, current gain of transistors, and/or anything else that can be measured electronically.
Multiplier Resistor	Resistor connected in series with a moving coil meter movement to extend the voltage ranges.
Multisegment Display	Device made of several light emitting diodes arranged in a numeric or alphanumeric pattern. By lighting selected segments numeric or alphabet characters can be displayed.
Multivibrator	A class of circuits designed to produce square waves or pulses. Astable multivibrators produce continous pulses without an external stimulus or trigger. Monostable multivibrators produce a single pulse for some predetermind period of time only when triggered. Bistable multivibrators produce a DC output that is stable in either one of two states, either high or low. An external stimulus or trigger is required for the bistable circuit to change states, either high to low or low to high.
Mutual Inductance	Ability of one inductor's lines of force to link with another inductor.
N-Type Semiconductor	A semiconductor compound formed by doping an intrinsic semiconductor with a pentavalent element. An N-type material contains an excess of conduction band electrons.

Term	Definition
Negative	Terminal that has an excess of electrons.
Negative Charge	A charge that has more electrons than protons.
Negative Feedback	A feedback signal 180° out of phase with an amplifier input signal. Used to increase amplifier stability, bandwidth, and input impedance. Also reduces distortion.
Negative Ground	A system where the negative terminal of the source is connected to the system's metal chassis.
Negative Ion	An atom having a greater number of electrons in orbit than there are protons in the nucleus.
Negative Resistance	A resistance such that when the current through it increases the voltage drop across the resistance decreases.
Negative Temperature Coefficient	A term used to describe a component whose resistance or capacitance decreases when temperature increases.
Neon Bulb	Glass envelope filled with neon gas which when ionized by an applied voltage will glow red.
Network	Combination of interconnected components, circuits, or systems.
Neutral	A terminal, point, or object with balanced charges. Neither positive nor negative.
Neutral Atom	An atom in which the number of negative charges (electrons in orbit) is equal to the number of positive charges (protons in the nucleus).
Neutral Wire	The conductor of a polyphase circuit or a single-phase, three-wire circuit that is intended to have a ground potential. The potential difference between the neutral and each of the other conductors are approximately equal in magnitude and equally spaced in phase.
Neutron	Subatomic particle in the nucleus of an atom and having no electrical charge.
Nickel-Cadmium Cell	A secondary cell that uses a nickel oxide positive electrode and a cadmium negative electrode.
Node	Junction or branch point in a circuit.
Noise	Unwanted electromagnetic radiation within an electrical or mechanical system.
Non-Inverting Input	An operational amplifier circuit having no phase inversion between the input and output. The terminal on an operational amplifier that is identified by a plus sign.
Non-Linear Scale	A scale in which the divisions are not equally spaced.
Normally Closed	Designation that states that the contacts of a switch or relay are closed or connected when at rest. When activated, the contacts open or separated.
Normally Open	Designation that states that the contacts of a switch or relay are normally open or not connected. When activated the contacts close or become connected.
North Pole	Pole of a magnet out of which magnetic lines of force are assumed to originate.
Norton's Theorem	Any network of voltage sources and resistors can be replaced by a single current source in parallel with a single resistor.

Term	Definition
Notch Filter	A filter that blocks a narrow band of frequencies and passes all frequencies above and below the band.
NPN **Transistor**	A bipolar junction transistor in which a *p*-type base element is sandwiched between an *n*-type emitter and an n-type collector.
Nucleus	Core of an atom. The nucleus contains both positive (protons) and neutral (neutrons) subatomic particles.
Octave	Interval between two sounds whose fundamental frequencies differ by a ratio of 2 to 1. 440 Hz. is one octave above 220 Hz.
Offset Null	An op-amp control pin used to eliminate the effects of internal component voltages on the output of the device.
OHM	Unit of resistance symbolized by the Greek capital letter omega (Ω).
Ohmmeter	Device used to measure electrical resistance.
Ohm's Law	Relationship among voltage, current, and resistance. Ohm's law states that current in a resistance varies in direct proportion to voltage applied and inversely proportional to resistance.
Ohms Per Volt	Refers to a value of ohms per volt of full scale defection for a moving coil meter movement. The number of ohms per volt is the reciprocal of the amount of current required to produce full scale deflection of the needle. A meter requiring 50 microamps for full-scale deflection has an internal resistance of 20 kΩ per volt. The higher the ohms per volt rating, the more sensitive the meter.
One-Shot	Monostable multivibrator.
Op-Amp	Abbreviation for operational amplifier.
Open Loop Gain	Gain of an amplifier when no feedback is present.
Open Loop Mode	An amplifier circuit having no means of comparing the output with the input (no feedback).
Operational Amplifier	A high-gain DC amplifier that has a high input impedance and a low output impedance. Op-amps are the most basic type of linear integrated circuits.
Oscillate	To produce a continuous output waveform without an input signal present.
Oscillator	An electronic circuit that produces a continuous output waveform with only DC applied.
Oscilloscope	An instrument used to display a signal graphically. Shows signal amplitude, period, and waveshape in addition to any DC voltage present. A multiple trace oscilloscope can show two or more waveforms at the same time for phase comparison and timing measurements.
Out of Phase	When the maximum and minimum points of two or more waveshapes do not occur at the same time.
Output	Terminal at which a component, circuit, or piece of equipment delivers current, voltage, or power.
Output Impedance	Impedance measured across the output terminals of a device without a load connected.

Term	Definition
Output Power	Amount of power a component, circuit, or system can deliver to a load.
Overload	Codition that occurs when the load is greater than the system was designed to handle (load resistance too small, load current too high). Overload results in waveform distortion and/or overheating.
Overload Protection	Protective device such as a fuse or circuit breaker that outomatically disconnects a load when current exceeds a predetermined value.
Paper Capacitor	Fixed capacitor using oiled or waxed paper as a dielectric.
Parallel	Circuit having two or more paths for current flow. Also called shunt.
Parallel Resonant Circuit	Circuit having an inductor and a capacitor in parallel with one another. Circuit offers a high impedance at resonant frequency. Sometimes called a "tank circuit."
Pass Band	The range of frequencies that will be passed and amplified by a tuned amplifier. Also the range of frequencies passed by a band-pass filter.
Passive Component	Component that does not amplify a signal. Resistors and capacitors are examples.
Passive Filter	A filter that contains only passive or nonamplifying components.
Passive System	System that emits no energy. It only receives. It does not transmit or reveal its position.
Peak	Maximum or highest amplitude level.
Peak Inverse Voltage (PIV)	The maximum rated value of a AC voltage acting in the direction opposite to that in which a device is designed to pass current.
Peak to Peak	Difference between the maximum positive and maximum negative values of an AC waveform.
Pentavalent Element	Element whose atoms have five valence electrons. Used in doping intrinsic silicon or germanium to produce n-type semiconductor material. Most commonly used pentavalent materials are arsenic and phosphorus.
Percent of Regulation	The change in output voltage that occurs between no-load and full-load in a DC voltage source. Dividing this change by the full-load value and multiplying the result by 100 gives percent regulation.
Percent of Ripple	The ratio of the effective rms value of ripple voltage to the average value of the total voltage. Expressed as a percentage.
Period	Time to complete one full cycle of a periodic or repeating waveform.
Permanence	Magnetic equivalent of magnetic inductance and consequently equal to the reciprocal of reluctance, just as conductance is equal to the reciprocal of resistance.
Permanent Magnet	Magnet normally made of hardened steel that retains its magnetism indefinately.
Permeability	Measure of how much better a material is as a path for magnetic lines of force with respect to air which has a permeability of one. Symbolized by the Greek lowercase letter mu (μ).

Term	Definition
Phase	Angular relationship between two waves.
Phase Angle	Phase difference between two or more waves, normally expressed in degrees.
Phase Shift	Change in phase of a wave form between two points, expressed as degrees of lead or lag.
Phase Shift Oscillator	An oscilator that uses three RC networks in its feedback path to produce the 180° phase shift required for oscillation.
Phase Splitter	Circuit that takes a single input signal and produces two output signals that are 180° apart in phase.
Phonograph	Piece of equipment used to reproduce sound stored on a disk called a phonograph record.
Phosphor	Luminescent material applied to the inner face of a cathode ray tube that when bombarded with electrons will emit light of various colors.
Photoconductive Cell	Material whose resistance decreases or conductance increases when exposed to light.
Photoconduction	A process by which the conductance of a material is change by incident electromagnetic radiation in the visible light spectrum.
Photodetector	Component used to detect or sense light.
Photodiode	A semiconductor diode that changes its electrical characteristics in response to illumination.
Photometer	Meter used to measure light intensity.
Photon	Discrete portion of electromagnetic energy. A small packet of light.
Photoresistor	Also known as a photoconductive cell or light-dependent resistor (LDR). A device whose resistance decreases with exposure to light.
Photovoltaic Cell	Component commonly called a solar cell used to convert light energy into electrical energy.
Pi	Value representin the ratio between the circumference and diameter of a circle and equal to approximately 3.142. Symbolized by the Greek lowercase letter π.
Pierce Oscillator	A variation of the colpitts oscillator. This oscillator uses a quartz crystal in place of the inductor found in the colpitts oscillator feedback network. The crystal maintains a highly stable output frequency.
Piezoelectric Crystal	Crystal material that will generate a voltage when mechanical pressure is applied and conversely will undergo mechanical stress when subjected to a voltage.
Piezoelectric Effect	The production of a voltage between opposite sides of a piezoelectric crystal as a result of pressure or twisting. Also the reverse effect which the application of a voltage to opposite sides causes a deformation to occur at the frequency of the applied voltage. (Converts mechanical energy into electrical energy and electrical energy into mechanical energy.)
Pinch-Off Region	A region on the characteristic curve of an FET in which the gate bias causes the depletion region to extend completely across the channel.

Term	Definition
Plastic Film Capacitor	Capacitor in which alternate layers of aluminum foil are separated by thin films of plastic dielectric.
Plate	Conductive electrode in either a capacitor or battery. In vacuum tube technology, it is the name given to the anode.
Plug	Movable connector that is normally connected into a socket or jack.
***PNP* Transistor**	A bipolar junction transistor with an n-type base and p-type emitter and collector.
Pole	In an active filter, a single RC circuit. A one pole filter has one capacitor and one resistor. A two pole filter has two RC circuits and so on.
Polar Coordinates	Either of two numbers that locate a point in a plane by its distance from a fixed point and the angle this line makes with a fixed line.
Polarity	Term used to describe positive and negative charges.
Polarized	A component that must be connected in correct polarity to function and/or prevent destruction. Example: Electrolytic capacitor.
Positive	Polarity of point that attracts electrons as opposed to negative that supplies electrons.
Positive Charge	A charge that exists in a body that has fewer electrons than protons.
Positive Feedback	A feedback signal that is in phase with an amplifier input signal. Positive feedback is necessary for oscillation to occur.
Positive Ground	A system whereby the positive terminal of the source is connected to the system's conducting chassis.
Positive Ion	Atom that has lost one or more valence electrons resulting in a net positive charge.
Potential Difference	Voltage difference between two points which will cause current to flow in a closed circuit.
Potential Energy	Energy that has potential to do work because of its position relative to others.
Potentiometer	A variable resistor with three terminals. Mechanical turning of a shaft can be used to produce variable resistance and potential. Example: A volume control is usually a potentiometer.
Power	Amount of energy converted by a circuit or component in a unit of time, normally seconds. Measured in units of watts (joules/second).
Power Amplifier	An amplifier designed to deliver maximum power output to a load. Example: In an audio system, it is the power amplifier that drives the loudspeaker.
Power Derating Factor	A transistor rating that tells how much the maximum allowable value of P_D decreased for each 1°C rise in ambient temperature.
Power Dissipation	Amount of heat energy generated by a device in 1 second when current flows through it.
Power Factor	Ratio of actual power to apparent power.

Term	Definition
Power Loss	Ratio of power absorbed to power delivered.
Power Supply	Electrical equipment used to deliver either AC or DC voltage.
Power Supply Rejection Ratio	A measure of an op-amp's ability to maintain a constant output when the supply voltage varies.
Primary	First winding of a transformer. Winding that is connected to the source as opposed to secondary, which is a winding connected to a load.
Primary Cell	Cell that produces electrical energy through an internal electrochemical action. Once discharged a primary cell cannot be reused.
Printed Circuit Board	Insulating board containing conductive tracks for circuit connections.
Programmable UJT	Unijunction transistor with a variable intrinsic stand-off ratio.
Propagation	Traveling of electromagnetic, electrical, or sound waves through a medium.
Propagation Delay	Time required for a signal to pass through a device or circuit.
Propagation Time	Time required for a wave to travel between two points.
Protoboard	Board with provision for attatching components without solder. Also called a breadboard. Primarily used for constructing experimental circuits.
Proton	Sub atomic particle within the nucleus of an atom. Has a positive charge.
Pulse	Rise and fall of some quantity (usually voltage) for a period of time.
Pulse Fall Time	Time for a pulse to decrease from 90% of its peak value to 10% of its peak value.
Pulse Repetition Frequency	The number of times per second that a pulse is transmitted. Pulse rate.
Pulse Repetition Time	Time interval between the start of two consecutive pulses.
Pulse Rise Time	Time required for a pulse to increase from 10% of its peak value to 90% of its peak value.
Pulse Width	Time interval between the leading edge and trailing edge of a pulse at a point where the amplitude is 50% of the peak value.
Push-Pull Amplifier	Amplifier using two active devices operating 180° apart.
Pythagorean Theorem	A theorem in geometry: The square of the hypotenuse of a right triangle equals the sum of the squares of the other two sides. In electronics used for vector analysis of AC circuits.
Q	Quality factor of an inductor or capacitor. It is the ratio of a component's reactance (energy stored) to its effective series resistance (energy dissipated). For a tuned circuit, a figure of merit used in bandwidth calculations. Q is the ratio of reactive power to resistive power in a tuned circuit. Also the symbol for charge in coulombs (Q for quantity).
Quiescent	At rest. For an amplifier the term is used to describe a condition with no active input signal.
Quiescent Point	(Q point) A point on the DC load line of a given amplifier that represents the quiescent (no signal) value of output voltage and current for the circuit.

Term	Definition
Radar	Acronym for "radio detection and ranging." A system that measures the distance and direction of objects.
Radioastronomy	Branch of astronomy that studies the radio waves generated by celestial bodies and uses these emissions to obtain information about them.
Radio Broadcast	Transmission of music, voice, and other information on radio carrier waves that can be received by the general public.
Radiocommunication	Term used to describe the transfer of information between two or more points by use of radio or electromagnetic waves.
Radio-Frequency Amplifier	Amplifier having one or more active devices to amplify radio signals.
Radio-Frequency Generator	Generator capable of supplying RF energy at any desired frequency in the radio-frequency spectrum.
Radio-Frequency Probe	Probe used in conjunction with an AC meter to measure radio-frequency signals.
RC	Abbreviation for "resistance capacitance," also abbreviation for "radio controled" as in "RC model airplanes."
RC Time Constant	Product of resistance and capacitance in seconds.
Reactance	Symbol "X." Opposition to current flow without the dissipation of energy. Example: The opposition provided by inductance or capacitance to AC current.
Reactive Power	Also called imaginary power or wattless power. It is the power value in "volt amps" obtained from the product of source voltage and source current in a reactive circuit.
Real Number	Number having no imaginary part.
Receiver	Unit or piece of equipment used to receive information.
Recombination	Process by which a conduction band electron gives up energy (in the form of heat or light) and falls into a valence band hole.
Rectangular Coordinates	A Cartesian coordinate of a Cartesion coordinate system whose straight-line axes or coordinate planes are perpendicular.
Rectangular Wave	Also known as a pulse wave. A repeating wave that only operates between two levels or values and remains at one of these values for a small amount of time relative to the other value.
Rectification	Process that converts alternating current to direct current.
Rectifier	Diode circuit that converts alternating current into pulsating direct current.
Reed Relay	Relay consisting of two thin magnetic strips within a glass envelope. When a coil around the envelope is energized, the relay's contacts snap together, making a connection between leads attached to the reed strips.
Regenerative Feedback	Positive feedback. Feedback from the output of an amplifier to the input such that the feedback signal is in phase with the input signal. Used to produce oscillation.

Term	Definition
Regulated Power Supply	Power supply that maintains a constant output voltage under changing load conditions.
Regulator	Device or circuit that maintains a desired output under changing conditions.
Relay	Electromechanical device that opens or closes contacts when a current is passed through a coil.
Relative	Not independent. Compared with or with respect to some other measured quantity.
Relaxation Oscillator	Free running circuit that outputs pulses with a period dependent or one or more RC time constants.
Reluctance	Resistance to the flow of magnetic lines of force.
Remanence	Amount a material remains magnetized after the magnetizing force has been removed.
Residual Magnetism	Magnetism remaining in the core of an electromagnet after the coil current is removed.
Resistance	Symbolized "R" and measured in ohms. Opposition to current flow and dissipation of energy in the form of heat.
Resistive Power	Amount of power dissipated as heat in a circuit containing resistive and reactive components. True power as opposed to reactive power.
Resistive Temperature Detector (RTD)	Temperature detector consisting of a fine coil of conducting wire (such as platinum) that will produce a relatively linear increase in resistance as temperature increases.
Resistivity	Measure of a material's resistance to current flow.
Resistor	Component made of material that opposes flow of current and therefore has some value of resistance.
Resistor Color Code	Coding system of colored stripes on a resistor to indicate the resistor's value and tolerance.
Resonance	Circuit condition that occurs at the frequency where inductive reactance (X_L) equals capacitive reactance (X_C).
Reverse Bias	Bias on a *PN* junction that allows only leakage current (minority carriers) to flow. Positive polarity on the *n*-type material and negative polarity to the *p*-type material.
Reverse Breakdown Voltage	Amount of reverse bias that will cause a *PN* junction to break down and conduct in the reverse direction.
Reverse Current	Current through a diode when reverse biased. An extremely small current also referred to as leakage.
Reverse Saturation Current	Reverse current through a diode caused by thermal activity. This current is not affected by the amount of reverse bias on the component, but does vary with temperature.
RF	Abbreviation for "radio frequency."

Term	Definition
Rheostat	Two terminal variable resistor used to control current.
Right Angle Triangle	Triangle having a 90° or square corner.
Ripple Frequency	Frequency of the ripple present in the output of a DC source.
Ripple Voltage	The small variations in Dc voltage that remain after filtering in a power supply.
Rise Time	Time for the leading edge of a pulse to rise from 10% of its peak value to 90% of its peak value.
RL Differentiator	An RL circuit whose output voltage is proportional to the rate of change of the input voltage.
RL Filter	Selective circuit of resistors and inductors that offers little or no opposition to certain frequencies while blocking or attenuating other frequencies.
RL Integrator	RL circuit with an output proportionate to the integral of the input signal.
RMS	Abbreviation for "root mean square."
RMS Value	The rms value of an AC sine wave is 0.707 times the peak value. This is the effective value of an AC sine wave. The rms value of a sine wave is the value of a DC voltage that would produce the same amount of heat in a heating element.
Roll-Off Rate	Rate of change in gain when an amplifier is operated outside of its bandwidth.
Rotary Switch	Electromechanical device that has a rotating shaft connected to one terminal capable of making or breaking a connection to one or more other terminals.
R-2R Ladder	Network or circuit composed of a sequence of L networks connected in tandem. Circuit used in digital to analog converters.
Saturation	Condition in which a further increase in one variable produces no further increase in the resultant effect. In a bipolar junction transistor, the condition when the emitter to collector voltage is less than the emitter to base voltage. This condition puts forward bias on the base to collector junction.
Sawtooth Wave	Repeating waveform that rises from zero to maximum value linearly drops back to zero and repeats. A ramp waveform.
Scale	Set of markings used for measurement.
Schematic Diagram	Illustration of an electrical or electronic circuit with the components represented by their symbols.
Schmitt Trigger	Circuit to convert a given waveform to a square wave output.
Schottky Diode	High-speed diode that has very little junction capacitance. Also known as a "hot-carrier diode" or a "surface-barrier diode."
Scientific Notation	Numbers entered as a number from one to ten multiplied by a power of 10. Example: $8765 = 8.765 \times 10^3$.
Secondary	Output winding of a transformer. Winding that is connected to a load.
Secondary Cell	Electrolytic cell used to store electricity. Once discharged may be restored by recharging by putting current through the cell in the direction opposite to that of discharge current.

Term	Definition
Selectivity	Characteristic of a circuit to discriminate between wanted and unwanted signals.
Self-Biasing	Gate bias for a field effect transistor in which source current through a resistor produces the voltage for gate to source bias.
Self-Inductance	Property that causes a counterelectromotive force to be produced in a conductor when the magnetic field expands or collapses with a change of current.
Semiconductor	An element that is neither a good conductor nor a good insulator, but rather lies somewhere between the two. Characterized by a valence shell containing four electrons. Silicon, germanium, and carbon are the semiconductors most frequently used in electronics.
Series Circuit	Circuit in which the components are connected end to end so that current has only one path to follow through the circuit.
Series pParallel Network	Network that contains components connected in both series and parallel.
Series Resonance	Condition that occurs in a series LC circuit at the frequency where inductive reactance equals capacitive reactance. Impedance is minimum; current is maximum limited only by resistance in the circuit.
Seven Segment Display	Device made of several light emitting diodes arranged in a numeric or alphanumeric pattern. By lighting selected segments numeric or alphabet characters can be displayed.
Shells or Bands	Orbital path containing a group of electrons having a common energy level.
Shield	Metal grounded cover used to protect a wire, component, or piece of equipment from stray magnetic and/or electric fields.
Short Circuit	Also called a short. Low resistance conection between two points in a circuit typically causing excessive current.
Shunt Resistor	Resistor connected in parallel or in shunt with another component or circuit.
Signal	Electrical quantity that conveys information.
Signal to Noise Ratio	Ratio of the magnitude of the signal to the magnitude of noise usually expressed in decibels.
Silicon (Si)	Nonmetalic element (atomic number 14) used in pure form as a semiconductor.
Silicon-Controlled Rectifier (SCR)	Three terminal active device that acts as a gated diode. The gate terminal is used to turn the device on allowing current to pass from cathode to anode.
Silicon Controlled Switch	An SCR with an added terminal called an anode gate. A positive pulse either at the anode gate or the cathode gate will turn the device on.
Silicon Dioxide	Glasslike material used as the gate insulating material in a MOSFET.
Silicon Transistor	A bipolar junction transistor using silicon as the semiconducting material.
Silver (Ag)	Precious metal that does not easily corrode and is more conductive than copper.
Silver Mica Capacitor	Mica capacitor with silver deposited directly onto the mica sheets instead of using conductive metal foil.

Term	Definition
Silver Solder	Solder composed of silver, copper, and zinc. Has a melting point lower than pure silver but higher than lead-tin solder.
Simplex	Communication in only one direction at a time. Example: FAX.
Simulcast	Broadcasting a program simultaneously in two different forms; for example, a program broadcast in both AM and FM.
Sine	Sine of an angle of a right angle triangle is equal to the opposite side divided by the hypotenuse.
Sine Wave	Wave whose amplitude is the sine of a linear function of time. It is plotted on a graph that plots amplitude against time or radial degrees relative to the angular rotation of an alternator.
Single In-Line Package	Package containing several electronic components (generally resistors) with a single row of connecting pins.
Single Pole Double Throw (SPDT)	Three-terminal switch in which one terminal can be connected to either one of the other terminals.
Single Pole Single Throw (SPST)	Two-terminal switch or relay thet can open or close one circuit.
Single Sideband (SSB)	AM radio communication technique in which the transmitter suppresses one sideband and therefore transmits only a single sideband.
Single Throw Switch	Switch containing only one set of contacts that can be either opened or closed.
Sink	Device such as a load that consumes power or conducts away heat.
Sintering	Process of bonding either a metal or powder by cold pressing it into a desired shape and then heating to form a strong cohesive body.
Sinusoidal	Varying in proportion to the sine of an angle or time function. AC voltage in which the instantaneous value is equal to the sine of the phase angle times the peak value.
SIP	Abbreviation for "single in-line package."
Skin Effect	Tendency of high-frequency (RF) currents to flow near the surface layer of a conductor.
Slew Rate	The maximum rate at which the output voltage of an op-amp can change.
Slide Switch	Switch having a sliding button, bar, or knob.
Slow-Acting Relay	Slow-operating relay that when energized may not pull up the armature for several seconds.
Slow-Blow Fuse	Fuse that can withstand a heavy current (up to ten times its rated value) for a small period of time before it opens.
Snap Switch	Switch containing a spring under tension or compression that causes the contacts to come together suddenly when activated.
SNR	Abbreviation for "signal to noise ratio."
Soft Magnetic Material	Ferromagnetic material that is easily demagnetized.

Term	Definition
Software	Program of instructions that directs the operation of a computer.
Solar Cell	Photovoltaic cell that converts light into electric energy. Especially useful as a power source for space vehicles.
Solder	Metallic alloy used to join two metal surfaces.
Soldering	Process of joining two metallic surfaces to make an electrical contact by melting solder (usually tin and lead) across them.
Soldering Iron	Tool with an internal heating element used to heat surfaces being soldered to the point where the solder becomes molten.
Solenoid	An air core coil. Equipped with a movable iron core the solenoid will produce motion. As a result of current through the coil the iron core is pulled into the center of the winding. When the coil is deenergized, a spring pulls the movable core away from the center of the winding. Mechanical devices connected to the movable core are made to move as a result of current through the coil. Example: Electric door locks on some automobiles.
Solid Conductor	Conductor having a single solid wire instead of strands of fine wire twisted together.
Solid State	Pertaining to circuits where signals pass through solid semiconductor material such as transistors and diodes as opposed to vacuum tubes where signals pass through a vacuum.
Sonar	Acronym for "sound navigation and ranging." A system using reflected sound waves to determine the position of some target.
Sonic	Pertaining to sound.
Sound Wave	Pressure waves propagated through air or other plastic media. Sound waves are generally audible to the human ear if the frequency is between approximately 20 and 20,000 vibrations per second (hertz).
Source	Device that provides signal power or energy to a load.
Source Follower	FET amplifier in which signal is applied between gate and drain with output taken between source and drain. Also called "common drain."
Source Impedance	Impedance through which output current is taken from a source.
South Pole	Pole of a magnet into which magnetic lines of force are assumed to enter.
Spark	Momentary discharge of electrical energy due to ionization of air or other dielectric material separating two charges.
SPDT	Single pole double throw.
Speaker	Also called "loudspeaker." Transducer that converts electrical energy into mechanical energy at audio frequencies.
Spectrum	Arrangement or display of light or other forms of electromagnetic radiation separated according to wavelength, energy, or some other property.
Spectrum Analyzer	Instrument used to display the frequency domain of a waveform plotting amplitude against frequency.

Term	Definition
Speed-Up Capacitor	Capacitor added to the base circuit of a BJT switching circuit to improve the switching time of the device.
SPST	Abbreviation for "single pole single throw."
Square Wave	Wave that alternates between two fixed values for an equal amount of time.
Static	Crackling noise heard on AM radio receivers. Caused by electric storms or electric devices.
Static Electricity	Stationary electric charges.
Static Reverse Current	Reverse current through a zener diode when the reverse voltage across the diode is less than the zener voltage rating of the device.
Stator	Stationary part of some rotary device such as a variable capacitor.
Step-Down Transformer	Transformer in which the output AC voltage is less than the input AC voltage.
Step-Up Transformer	Transformer in which the output AC voltage is greater than the input AC voltage.
Stereo Sound	System in which reproduced sound is delivered through two or more channels to give a sense of direction to the source.
Stop Band	Range of frequencies outside the pass band of a tuned amplifier.
Storage Time	In a BJT switching circuit, it is the time required for collector current to drop from 100% to 90% of its maximum value.
Stranded Conductor	Conductor composed of a group of strands of wire twisted together.
Stray Capacitance	Undesirable capacitance that exists between two conductors such as two leads or one lead and a metal chassis.
Subassembly	Components contained in a unit for convenience in assembling or servicing equipment.
Subatomic	Particles such as electrons, protons, and neutrons that are smaller than atoms.
Substrate	Mechanical insulating support upon which a device is fabricated.
Summing Amplifier	An op-amp circuit whose output is proportional to the sum of its instantaneous voltages.
Superconductor	Metal such as lead or niobium that, when cooled to within a few degrees of absolute zero, can conduct current with no resistance.
Superheterodyne Receiver	Radio receiver that converts all radio frequencies to a fixed intermediate frequency to maximize gain and bandwidth before demodulation.
Super High Frequency (SHF)	Frequency band between 3 GHz and 30 GHz. So desiganted by Federal Communications Comission (FCC).
Superposition Theorem	Theorem designed to simplify networks containing two or more sources. It states that in a network containing more than one source, the current at any one point is equal to the algebraic sum of the currents produced by each source acting separately.

Term	Definition
Supply Voltage	Voltage provided by a power source.
Surface-Barrier Diode (Schottky diode)	High-speed diode that has very little junction capacitance. Also known as a "hot-carrier diode."
Surface Leakage Current	Diode reverse current that passes along the surface of the semiconductor materials.
Surge Current	High charging current that flows into a power supply filter capacitor as the power is first turned on.
Sweep Generator	Test instrument designed to produce a voltage that continously varies in frequency over a band of frequencies. Used as a souce to display frequency response of a circuit on an oscilloscope.
Switch	Electrical device having two states, on (closed) or off (open). Ideally having zero impedance when closed and infinite impedance when open.
Switching Transistor	transistor designed to change rapidly between saturation and cut-off.
Synchronization	Also called sync. Precise matching of two waves or functions.
Synchronous	Two or more signals in step or in phase.
Sync Pulse	Pulse used as a reference for synchronization.
System	Combination of several pieces of equipment to perform in a particular manner.
Tank Circuit	Parallel resonant circuit containing only a coil and a capacitor. Both the coil and capacitor store electrical energy for part of each cycle.
Tantalum Capacitor	Electrolytic capacitor having a tantalum foil anode. Able to have a large capacity in a small package.
Tap	Electrical connection to some point other than at the ends of a resistor or inductor.
Tapered	Nonunifrom distribution of resistance per unit length throughout the element of a potentiometer.
Technician	Expert in troubleshooting circuit and system malfunctions. Along with a thorough knowledge of test equipment and how to use it to diagnose problems, the technician is also familiar with how to repair or replace faulty components. Technicians basically translate theory into action.
Telegraphy	Communication between two points by sending and receiving a series of current pulses either through wire or by radio.
Telemetry	Transmission of instrument readings to a remote location either by wire or by radio.
Telephone	Apparatus designed to convert sound waves into electrical waves which are sent to and reproduced at a distant point.
Telephone Line	Wires existing between subscribers and central stations in a telephone system.
Telephony	Telecommunications system involving the transmission of speech information, allowing two or more persons to communicate verbally.

Term	Definition
Teletypewriter	Electric typewriter that like a teleprinter can produce coded signals corresponding to the keys pressed or print characters corresponding to the coded signals received.
Television	System that converts both audio and visual information into corresponding electrical signals that are then transmitted through wires or by radio waves to a receiver, which reproduces the original information.
Telex	Teletypewriter exchange service.
Temperature Coefficient of Frequency	Rate at which frequency changes with temperature.
Tera (T)	Metric prefixes that represents 10^{12}.
Terminal	Point at which electrical connections are made.
Tesla (T)	Unit of magnetic flux density (1 tesla = 1 Wb/m^2).
Test	Sequence of operations intended to verify the correct operation or malfunctioning of a piece of equipment or system.
Thermal Relay	Relay activated by a heating element.
Thermal Runaway	Problem that can develop in an amplifier when an increase in temperature causes an increase in collector current. The increase in collector current causes a further increase in temperature and so on. Unless the circuit is designed to prevent this condition, the device can be driven into saturation.
Thermal Stability	The ability of a circuit to maintain stable characteristics in spite of increased temperature.
Thermistor	Temperature-sensitive semiconductor that has a negative temperature coefficient of resistance. As temperature increases, resistance decreases.
Thermocouple	Temperature transducer consisting of two dissimilar metals welded together at one end to form a junction that when heated will generate a voltage.
Thermometry	Relating to the measuring of temperature.
Thermostat	Device that opens or closes a circuit in response to changes in temperature.
Thevenin's Theorem	Theorem that replaces any complex network with a single voltage source in series with a single resistance.
Thick Film Capacitor	Capacitor consisting of two thick-film layers of conductive film separated by a deposited thick-layer dielectric film.
Thick Film Resistor	Fixed value resistor consisting of thick-film resistive element made from metal particles and glass powder.
Thin Film Capacitor	Capacitor in which both the electrodes and the dielectric are deposited in layers on a substrate.
Thin Film Detector (TFD)	A temperature detector containing a thin layer of platinum and used for precise temperature readings.

Term	Definition
Three-Phase Supply	AC supply that consists of three AC voltages 120° out of phase with each other.
Threshold	Minimum point at which an effect is produced or detected.
Threshold Voltage	For an enhancement MOSFET, the minimum gate source voltage required for conduction of source drain current.
Thyristor	A term used to classify all four-layer semiconductor devices. SCRs and triacs are examples of thyristors.
Time Constant (τ)	Time required for a capacitor in an RC circuit to charge to 63% of the remaining potential across the circuit. Also time required for current to reach 63% of maximum value in an RL circuit. Time constant of an RC circuit is the product of R and C. Time constant of an RL circuit is equal to inductance divided by resistance.
Time Division Multiplexing (TDM)	Transmission of two or more signals on the same path, but at different times.
Time Domain Analysis	A method of representing a waveform by plotting amplitude over time.
Toggle Switch	Spring loaded switch that is put in one of two positions either on or off.
Tolerance	Permissible deviation from a specified value normally expressed as a percentage.
TO Package	Cylindrical, metal can type of package of some semiconductor components.
Toroidal Coil	Coil wound on a doughnut shaped core.
Transconductance	Also called mutual conductance. Ratio of a change in output current to the change in input voltage that caused it.
Transducer	Device that converts energy from one form to another.
Transformer	Inductor with two or more windings. Through mutual inductance, current in one winding called a primary will induce current into the other windings called secondaries.
Transformer Coupling	Also called inductive coupling. Coupling of two circuits by means of mutual inductance provided by a transformer.
Transistor	Term derived from "transfer resistor." Semiconductor device that can be used as an amplifier or as an electronic switch.
Transmission	Sending of information.
Transmission Line	Conducting line used to transmit signal energy between two points.
Transmitter	Equipment used to achieve transmission.
Triac	Bidirectional gate controlled thyristor similar to an SCR but capable of conducting in both directions. Provides full-wave control of AC power.
Triangular Wave	A repeating wave that has equal positive going and negative going ramps. The ramps have linear rates of change with time.
Trigger	Pulse used to initiate a circuit action.

Term	Definition
Triggering	Initiation of an action in a circuit, which then functions for a predetermined time. Example: The duration of one sweep in a cathode ray tube.
Trimmer	Small value variable capacitor, resistor, or inductor used to fine-tune a larger value.
Trivalent Element	One having three valence electrons. Used as an impurity in semiconductor material to produce p-type material. Most commonly used trivalent elements are aluminum, gallium, and boron.
Troubleshooting	Systematic approach to locating the cause of a fault in an electronic circuit or system.
Tuned Circuit	Circuit that can have its component values adjusted so that it responds to one selected frequency and rejects all others.
Tunnel Diode	Heavily doped junction diode that has negative resistance in the forward direction of its operating range.
Turn-Off Time	Sum of storage time and fall time.
Turn-On Time	Sum of delay time and rise time.
Turns Ratio	Ratio of the number of turns in the secondary winding of a transformer to the number of turns in the primary winding.
Two Phase	Two repeating waveforms having a phase difference of 90°.
UHF	Abbreviation for "ultra high frequency."
Ultrasonic	Signals that are just above the frequency range of human hearing of approximately 20 kHz.
Uncharged	Material having atoms with the same number of electrons in orbit as the number of protons in the nucleus. Having no electrical charge.
Unijunction Transistor	Three terminal device that acts as a diode with its own internal voltage divider biasing circuit.
Unity Gain Frequency	Frequency of operation for a device where the gain of the component drops to unity.
VA	Abbreviation for "volt ampere."
Vacuum Tube	Electron tube evacuated to such a degree that its electrical characteristics are essentially unaffected by the presence of residual gas or vapor. Have been essentially replaced by transistors for amplification and rectification. Cathode ray tubes are still used as display devices.
Valence Shell	The outermost electron shell for a given atom. The number of electrons in this shell determines the conductivity of the atom.
Varactor Diode	*PN* junction diode with a high junction capacitance when reverse biased. Most often used as a voltage controlled capacitor. The varactor is also called vari-cap, tuning diode, and epi-cap.
Variable Capacitor	Capacitor whose capacitance can be change by varying the effective area of the plates or the distance between the plates.

Term	Definition
Variable Resistor	Resistor whose resistance can be changed by turning a shaft. *See also Potentiometer* and *Rheostat.*
VCR	Abbreviation for "video cassette recorder."
Vector	Quantity having both magnitude and direction. Normally represented by a line. Length of the line indicates magnitude and orientation indicates direction.
Vector Diagram	Arrangement of vectors showing phase relationships between two or more AC quantities of the same frequency.
Vertical MOS	Enhancement type MOSFET designed to handle much greater values of drain current than standard E-MOSFET.
Very High Frequency (VHF)	Electromagnetic frequency band from 30 MHz to 300 MHz.
Very Low Frequency (VLF)	Frequency band from 3 kHz to 30 kHz.
Video	Relating to any picture or visual information. From the Latin word meaning "I see."
Video Amplifier	Amplifier having one or more stages designed to amplify video signals.
Virtual Ground	Point in a circuit that is always at approximately ground potential. Often a ground for voltage, but not for current.
Voice Coil	Coil attached to the diaphragm of a moving coil loudspeaker. The coil is moved through an air gap between magnetic pole pieces.
Voice Synthesizer	Synthesizer that can simulate speech by stringing together phonemes.
Volt	Unit of potential difference or electromotive force. One volt is the potential difference needed to produce one ampere of current through a resistance of one ohm.
Voltage (V)	Term used to designate electrical pressure or force that causes current to flow.
Voltage Amplifier	Amplifier designed to build up signal voltage. By design amplifiers can have a large voltage gain or a large current gain or a large power gain. Voltage amplifiers are designed to maximize voltage gain often at the expense of current gain or power gain.
Voltage Controlled Oscillator	Oscillator whose output frequency depends on an input control voltage.
Voltage Divider	Fixed or variable series resistor network connected across a voltage to obtain a desired fraction of that voltage.
Voltage Divider Biasing	Biasing method used with amplifiers in which two series resistors connected across a source. The junction of the two biasing resistors provides correct bias voltage for the amplifier.
Voltage Drop	Voltage or difference in potential developed across a component due to current flow.

Term	Definition
Voltage Feedback	Feedback configuration where a portion of the output voltage is fed back to the input of an amplifier.
Voltage Follower	Operational amplifier circuit characterized by a high input impedance, low output impedance, and unity voltage gain. Used as a buffer between a source and a low impedance load.
Voltage Gain	Also called voltage amplification. Ratio of amplifier output voltage to input voltage usually expressed in decibels.
Voltage Multiplier	Rectifier circuit using diodes and capacitors to produce a DC output voltage that is some multiple of the peak value of AC input voltage. Cost-effective way of producing higher DC voltages. Voltage doublers and voltage triplers are examples.
Voltage Rating	Maximum voltage a component can withstand without breaking down.
Voltage Regulator	Device or circuit that maintains constant output voltage (within certain limits) in spite of changing line voltage and/or load current.
Voltage Source	Circuit or device that supplies voltage to a load.
Voltaic Cell	Primary cell having two unlike electrodes immersed in a solution that chemically interacts to produce a voltage.
Volt-Ampere	Unit of apparent power in an AC circuit containing capacitive or inductive reactance. Apparent power is the sum of source voltage and current.
Voltmeter	Instrument used to measure difference in potential between two points.
Volume	Magnitude or power level of audio frequency. Measured in volume units (VU).
Watt	Unit of electrical power required to do work at the rate of one joule per second. One watt of power is expended when one ampere of direct current flows through a resistance of one ohm. In an AC circuit, true power is the product of effective volts and effective amperes, multiplied by the power factor.
Wattage Rating	Maximum power a device can safely handle continuously.
Watt-hour	Unit of electrical work, equal to a power of one watt being absorbed for one hour.
Wattmeter	Instrument used to measure electric power in watts.
Wave	Electric, electromagnetic, acoustic, mechanical, or other form whose physical activity rises and falls or advances and retreats periodically as it travels through some medium.
Waveform	Shape of a wave.
Waveguide	Rectangular or circular pipe used to guide electromagnetic waves at micro frequencies.
Wavelength (λ)	Distance between two points of corresponding phase and is equal to waveform velocity divided by frequency.

Term	Definition
Weber (Wb)	Unit of magnetic flux. One Weber is the amount of flux that when linked with a single turn of wire for an interval of one second will induce an electromotive force of one volt.
Wienbridge Oscillator	Oscillator that uses an RC low-pass filter and an RC high-pass filter to set the frequency of oscillations.
Wet Cell	Secondary cell using a liquid as an electrolyte.
Wetting	Term used in soldering to describe the condition that occurs when the metals are being soldered are hot enough to melt the solder so it flows over the surface.
Wheatstone Bridge	Four-arm bridge circuit used to measure resistance, inductance or capacitance.
Wideband Amplifier	Also called "broadband amplifier." Amplifier with a flat response over a wide range of frequencies.
Winding	One or more turns of a conductor wound in the form of a coil.
Wire	Single solid or stranded group of conductors having a low resistance to current flow. Used to make connections between circuits or points in a circuit.
Wire Gauge	American wire gauge (AWG) is a system of numerical designations of wire diameters.
Wireless	Term describing radio communication that requires no wired between two communicating points.
Wire-Wound Resistor	Resistor in which the resistive element is a length of high-resistance wire or ribbon, usually nichrome, wound onto an insulating form.
Wire Wrapping	Method of making a connection by wrapping wire around a rectangular pin.
Woofer	Large loudspeaker designed primarily to reproduce low-frequency audio signals.
Work	Work is done anytime energy is transformed from one type to another. The amount of work done is dependent on the amount of energy transformed.
X	Symbol for reactance.
x-axis	Horizontal axis.
Y	Symbol for admittance.
y-axis	Vertical axis.
z-axis	Axis perpendicular to both x- and y-axes.
Zener diode	Semiconductor diode in which reverse breakdown voltage current causes the diode to develop a constant voltage. Used as a clamp for voltage regulation.
Zeroing	Calibrating a meter so that it shows a value of zero when zero is being measured.

Electronic Acronyms

Term	Definition
A (amp)	Ampere
AC	Alternating current
AC/DC	Alternating current or direct current
A/D	Analog to digital
ADC	Analog-to-digital converter
AF	Audio frequency
AFT	Automatic fine-tuning
AFC	Automatic frequency control
AFC	Automatic flow controller, used in controlling the flow of gasses under pressure into a vacuum system
AGC	Automatic gain control
Ah	Ampere-hour

Electronic Acronyms courtesy Twisted Pair (*www.twysted-pair.com*)

Term	Definition
A_i	Current gain
AM	Amplitude modulation
AM/FM	Amplitude modulation or Frequency modulation
AMM	Analog multi-meter
Antilog	Antilogarithm
A_p	Power gain
apc	Automatic phase control
A_v	Voltage gain
AVC	Automatic volume control
AWG	American wire gauge
B	Flux density
BCD	Binary coded decimal
Bfo	Beat frequency oscillator
BJT	Bipolar junction transistor
BW	Bandwidth
c	Centi (10^{-2})
C	Capacitance or capacitor
CAD	Computer aided design
CAM	Computer aided manufacture
CATV	Cable TV
CB	Common base configuration
CB	Citizen's band
CC	Common collector
CE	Common emitter
cm	Centimeter
cmil	Circular mil
CPU	Central processing unit
C (Q)	Coulomb
CR cr	Junction diode
CRO	Cathode ray oscilloscope
CRT	Cathode ray tube
C_T	Total capacitance

Term	Definition
cw	Continuous transmission
d	Deci (10^{-1})
D/A or D-A	Digital to analog
DC	Direct current
Di or Δi	Change in current
DIP	Dual in-line package
DMM	Digital multimeter
DPDT	Double pole double throw
Dt or Δt	Change in time
DTL	Diode transistor logic
Dv or Δv	Change in voltage
DVM	Digital voltmeter
E DC or Erms	Difference in potential
e	Instantaneous difference in potential
ECG	Electrocardiogram
ECL	Emitter coupled logic
EHF	Extremely high frequency
EHV	Extra high voltage
ELF	Extremely low frequency
EMF	Electromotive force
EMI	Electromagnetic interference
EW	Electronic warfare
f	Frequency
FET	Field effect transistor
FF	Flip flop
fil	Filament
FM	Frequncy modulation
f_r	Frequency at resonance
fsk	Frequency-shift keying
FSD	Full-scale deflection
G	Gravitational force
G	Conductance

Term	Definition
G	Giga (10^9)
H	Henry
H	Magnetic field intensity
H	Magnetizing flux
h	hecto (10^2)
h	Hybrid
HF	High frequency
hp	Horsepower
Hz	Hertz
I	Current
i	Instantaneous current
I_B	DC base current
I_C	DC collector current
IC	Integrated circuit
I_e	Total emitter current
I_{eff}	Effective current
IF	Intermediate frequency
I_{max}	Maximum current
I_{min}	Minimum current
I/O	Input/output
IR	Infrared
I_R	Resistor current
I_S	Secondary current
I_T	Total current
JFET	Junction field effect transistor
K	Coefficient of coupling
k	Kilo (10^3)
kHz	Kilohertz
kV	Kilovolt
kVA	Kilovoltampere
kW	Kilowatt
kWh	Kilowatt-hour

Term	Definition
L	Coil, inductance
LC	Inductance-capacitance
LCD	Liquid crystal display
L-C-R	Inductance-capacitance-resistance
LDR	Light-dependent resistor
LED	Light emitting diode
LF	Low frequency
L_M	Mutual inductance
LNA	Low noise amplifier
LO	Local oscillator
LSI	Large scale integration
L_T	Total inductance
M	Mega (10^6)
M	Mutual conductance
M	Mutual inductance
m	Milli (10^{-3})
mA	Milliampere
mag	Magnetron
max	Maximum
MF	Medium frequency
mH	Millihenry
MHz	Megahertz
min	Minimum
mm	Millimeter
mmf	Magnetomotive force
MOS	Metal oxide semiconductor
MOSFET	Metal oxide semiconductor field effect transistor
MPU	Microprocessor unit
MSI	Medium-scale integrated circuit
mV	Millivolt
mW	Milliwatt
N	Number of turns in an inductor

Term	Definition
N	Revolutions per minute
n	Nano (10^{-9})
N	Negative
nA	Nanoampere
NC	Normally closed
NC	No connection
NEG, neg	Negative
nF	Nanofarad
nH	Nanohenry
nm	Nanometer
NO	Normally open
NPN	Negative-positive-negative
ns	Nanosecond
nW	Nanowatt
Op-Amp	Operational amplifier
P	Pico (10^{-12})
P	Power
p	Instantaneous power
P	Positive, also peak
PA	Public address or power amplifier
pA	Picoampere
PAL	Programmable Array Logic
PAM, pam	Pulse amplitude modulation
P_{ap}	Apparent power
P_{av}	Average power
PCB	Printed circuit board
PCM, pcm	Pulse-code modulation
PDM	Pulse-duration modulation
pF	Picofarad
PLD	Programmable logic device
PLL	Phase-locked loop
PM	Phase modulation, also Permanent magnet

Term	Definition
PNP	Positive-negative-positive
POT, pot	Potentiometer
P-P	Peak to peak
PPM	Pulse-position modulation
PRF	Pulse repetition frequency
PRT	Pulse repetition time
pw	Pulse width
PWM, pwm	Pulse width modulation
Q	Charge, also quality
q	Instantaneous charge
R	Potentiometer
R	Resistance
RAM	Random access memory
RC	Resistance-capacitance, also Radio controlled
rcvr	Receiver
rect	Rectifier
ref	Reference
rf	Radio frequencies
RF	Radio frequencies
RFI	Radio frequency interference
R_L	Load resistor
RLC	Resistance-capacitance-inductance
RMS, rms	Root mean square
ROM	Read only memory
rpm	Revolutions per minute
SCR	Silicon controlled rectifier
SHF	Super high frequency
SIP	Single in-line package
SNR	Signal-to-noise ratio
SPDT	Single pole double throw
sq cm	Square centimeter
SSB	Single sideband

Term	Definition
SW	Short wave
SWR	Standing-wave ratio
SYNC, sync	Synchronous
T	Tera (10^{12})
T	Torque
T	Transformer
t	Time in seconds
TC	Time constant, also temperature coefficient
TE	Transverse electric
temp	Temperature
THz	Terahertz
TM	Transverse magnetic
TR	Transmit-receive
TTL	Transistor-transistor logic
TV	Television
TWT	Travelling wave tube
UHF	Ultra high frequency
UHV	Ultra high voltage
UJT	Unijunction transistor
UV	Ultraviolet
V	Vacuum tube
V, v	Volt
v	Instantaneous voltage
VA	Volt ampere
V_{av}	Voltage (average value)
V_{BE}	DC voltage base to emitter
V_c	Capacitive voltage
V_{CE}	DC voltage collector to emitter
VCO	Voltage controlled oscillator
VHF	Very high frequency
V_{in}	Input voltage
V_L	Inductive voltage

Term	Definition
VLF	Very low frequency
V_m, V_{max}	Maximum voltage
VOM	Volt ohm milliameter
V_{out}	Output voltage
V_p	Primary voltage
V_S	Source voltage
VSWR	Voltage standing wave ratio
V_T	Total voltage
W	Watt
$\mathbf{X_C}$	Capacitive reactance
$\mathbf{X_L}$	Inductive reactance
Y	Admittance
Z	Impedance
$\mathbf{Z_{in}}$	Input impedance
$\mathbf{Z_o}$	Output impedance
$\mathbf{Z_p}$	Primary impedance
$\mathbf{Z_s}$	Secondary impedance
$\mathbf{Z_T}$	Total impedance
kΩ	Kilohm
MΩ	Megohm
μA	Microamp
μF	Microfarad
μH	Microhenry
μV	Microvolt
μW	Microwatt
Ω	Ohm
°	Degrees
°C	Degrees centigrade or Celcius
°F	Degrees Farenheit

PCB Design Acronyms

Term	Definition
AA	Automatic acknowledge
AA	Antenna array
AGR	Annual average growth rate
ABS	Acrylonitrile-butadiene-styrene (plastic)
AC	Alternating current
AC	All call
ACA	Anisotropically conductive adhesive
ACC	Advanced concept construction active control channel
ACCU	Alternating current connection unit
AEC	Architecture, engineering, and construction
Agc	Automatic gain control
AGV	Automated guided vehicle

Industry acronyms courtesy IPC (www.IPC.org)

Term	Definition
AI	Artificial intelligence
AIS	Adhesive interconnect system
ANOVA	Analysis of variance
ANSI	American National Standards Institute
AOI	Automated optical inspection
AOQ	Average outgoing quality
APT	Automatically programmed tools
AQL	Acceptable quality level
ARINC	Aeronautical Radio Incorporated
ASCII	American Standard Code for Information Interchange
ASIC	Application specific integrated circuit
ASME	American Society of Mechanical Engineers
ASPaRC	Ability of solder paste to retain components
ASQ	American Society for Quality
ASTM	American Society for Testing and Materials
ATE	Automatic test equipment
ATG	Automatic test generation
ATR	Air transport rack
AVT	Accelerated vesication test
AWG	American wire gauge
AXI	Automated x-ray inspection
BC	Buried capacitance
BDMA	Benzyldimethylamine
BGA	Ball grid array
BITE	Built-in test equipment
BOD	Biochemical oxygen demand
BOM	Bill of material
BOT	Build to order
BT	Bismaleimide triazine
BTAB	Bumped tape—automated bonding
C3	Command, control, and communicate
C4	Controlled collapse component connection

Term	Definition
CAD	Computer-aided design
CAE	Computer-aided engineering
CAF	Cathotic anionic filaments
CAFM	Computer-aided facilities management
CAGE	Commercial and government entity
CALS	Computer-aided acquisition and logistic support (DOD)
CAM	Computer-aided manufacturing
CAPP	Computer-aided process planning
CAR	Computer-aided repair
CASE	Computer-aided software engineering
CAT	Computer-aided testing
CBGA	Ceramic ball grid array
CCAPS	Circuit card assembly and processing system
CDA	Copper development association
C&E	Cause and effect
CET	Certified electronic technician
CEPM	Certified EMS program managers
CFM	Continuous flow manufacturing
CFM	Cubic feet per minute
CIM	Computer-integrated manufacturing
CISC	Complex instruction set computing
CITIS	Contractor integrated technical information services
CMC	Copper moly copper
CMOS	Complimentary metal-oxide semiconductor
CNC	Computer numerical control
COB	Chip-on-board
COD	Consumed oxygen demand
COT	Configure to order
Cp	Capability performance
CPL	Capability performance, lower
CPLD	Complex programmable logic device
CPU	Capability performance, upper

Term	Definition
CPU	Central processing unit (computer)
CRC	Cyclic redundancy check
CRT	Cathode ray tube
CSA	Canadian Standards Agency
CSG	Constructive solids geometry
CSP	Chip scale package
CTE	Coefficient of thermal expansion
CVS	Cyclic voltammetry stripping
DAB	Designated audit body
DATC	Design Automation Technical Committee (IEEE)
DBMS	Database management system
DC	Direct current
DCAS	Defense Contract Administration Service
DCMA	Defense Contract Management Agency
DCMC	Defense Contract Management Command
DESC	Defense electronics supply center
DFM	Design for manufacture
DIM	Data-information record
DIN	Deutsches institute for normung
DIP	Dual-inline package
Dk	Dielectric constant
DLA	Defense Logistics Agency (DOD)
DMA	Direct memory access
DMS	Dynamic mechanical spectroscopy
DMSA	Defense Manufacturers and Suppliers Association
DNC	Distributed (or direct) numerical control
DOD	Department of Defense
DOD	Dissolved oxygen demand
DOE	Design of experiments
DOS	Disk operating system
DRC	Design rule checking
DRM	Drawing requirements manual

Term	Definition
DS	Double sided
DSC	Differential scanning calorimetry
DSP	Digital signal processor
DTP	Diameter true position
DTS	Dock to stock
DVM	Digital voltmeter
DXF	Data exchange format
ECAD	Electronic computer-aided design
ECC	Error correction code
ECCB	Electronic Components Certification Board
ECL	Emitter-coupled logic
ECM	Electronic countermeasures
ECN	Engineering change notice
ECO	Engineering change order
ECR	Engineering change request
ED	Electrodeposited
EDA	Electronic design automation
EDI	Electronic data interchange
EDIF	Electronic design interchange format
EDM	Electro-discharge machining
EDO	Extended data out
EIA	Electronics Industry Association
EIS	Engineering information system
ELD	Electroluminescent diode
EMC	Electromagnetic compatibility
EMF	Electromotive force
EMI	Electromagnetic interference
EMP	Electromagnetic pulse
EMPF	Electronics manufacturing productivity facility
EPA	Environmental Protection Agency
EPR	Ethylene-propylene (copolymer) resin
EPT	Ethylene-propylene terepolymer

Term	Definition
ESD	Electrostatic discharge
ESI	Early supplier involvement
ESR	Equivalent series resistance
ETPC	Electrolytic tough-pitch copper
FAA	Federal Aviation Administration
FAR	Failure analysis report
FC-CBGA	Flip chip ceramic ball grid array
FCC	Federal Communications Commission
FCC	Flat-conductor cable
FC-PBGA	Flip chip platic ball grid array
FEA	Finite-element analysis
FEM	Finite-element modeling
FEP	Fluorinated ethylene-propylene (Teflon)
FET	Field-effect transistor
FFT	Fast fourier transform
FMEA	Fault node and effect analysis
FPGA	Field programmable gate array
FPT	Fine-pitch technology
FSCM	Federal Stock Code for Manufacturers
FTP	File Transfer Protocol
GaAs	Gallium arsenide
GBIB	General purpose interface bus
GMA	Gas metal arc (welding)
GTA	Gas tungsten arc (welding)
GTPBGA	Glob top plastic ball grid array
HASL	Hot air solder level
HPGL	Hewlett Packard Graphic Language
HTE	High-temperature elongation
HTML	Hypertext Markup Language
HTTP	Hypertext Transfer Protocol
I/O	Input/output (terminations)
IC	Integrated circuit
ICA	Isotropically conductive adhesive

Term	Definition
ICAM	Integrated computer-aided manufacturing
IDC	Insulation-displacement connection
IEC	International Electrotechnical Commission
IECQ	International Electronic Component Qualification System
IEDR	Initial engineering design review
IEEE	Institute of Electrical and Electronic Engineers
IEPS	International Electronic Packaging Society
IGES	Integrated Graphics Exchange System
ILB	Inner-lead bonding (TAB)
IP	Internet Protocol
IPM	Inches per minute
IR	Infrared
ISCET	International Society of Certified Electronics Technician
ISHM	International Society for Hybrid Microelectronics
ISO	International Organization for Standardization
ITT	Inter-test time
JEDEC	Solid State Technology Association (Formerly the Joint Electronic Device Engineering Council)
JIT	Just-in-time (manufacturing)
KGB	Known good board
KGD	Known good die
LAN	Local area network
LBA	Logical block address
LCCC	Leadless ceramic chip carrier
LCD	Liquid crystal display
LDA	Logic design automation
LED	Light-emitting diode
LGA	Lang grid array
LIF	Low insertion force
LMC	Least material condition
LPISM	Liquid photo-imageable solder mask
LRU	Lowest replaceable unit
LSI	Large-scale integration (integrated circuit)

Term	Definition
LTCC	Low temperature co-fired ceramic
MA	Mechanical advantage
MAC	Maximum allowable concentration
MAC	Media access control
MAP	Manufacturing automation protocol
MATS	Material transport segment
MCAD	Mechanical computer-aided design
MCAE	Mechanical computer-aided engineering
MCM	Multichip module
MDA	Methylenedianiline
MEK	Methyl-ethyl ketone
MELF	Metal electrode face (discrete leadless component)
MIBK	Methyl-isobutyl ketone
MIR	Moisture insulation resistance
MITI	Ministry of International Trade and Industry (Japan)
MLB	Multilayer board
MLPWB	Multilayer printed wiring board
MMC	Maximum material condition
MMIC	Monolithic microwave integrated circuit
MOS	Metal-oxide semiconductor
MRP	Material requirement planning
MRP II	Manufacturing resource planning
MSDS	Material safety data sheets
MSI	Medium-scale integration (integrated circuit)
MTBF	Mean time between failures
MTTR	Mean time to repair
NADCAP	National Aerospace and Defense Contractors Accreditation Procedures
NASA	National Aeronautics and Space Administration
NBR	Nitrile butadiene-acrylonitrile rubber
NBS	National Bureau of Standards
NC	Numerical control
NDT	Nondestructive testing

Term	Definition
NECQ	National Electronics Component Qualification System
NEMA	National Electrtical Manufacturers Association
NIST	National Institute for Science and Technology
NMR	Normal-mode rejection
NSA	National Security Agency
OA	Organic acid (flux)
ODR	Oscillating disk rheometer
OEM	Original equipment manufacturer
OFHC	Oxygen-free high-conductivity copper
OLB	Outer-lead bonding (TAB)
OSHA	Occupational Safety Hazards Act
OSI	Open systems interconnection
OSP	Organic solder preservative
P&IA	Packaging and interconnecting assembly
P&IS	Packaging and interconnecting structure
PBGA	Plastic ball grid array
PBX	Private branch exchange
PC	Personal computer
PCA	Printed circuit assembly
PCB	Printed circuit board
PCMCIA	Personal Computer Memory Card International Association
PDES	Product data exchange specification
PDL	Page description language
PEM	Plastic electronic module
PGA	Pin grid array (leaded component package)
PHIGS	Programmer's Hierarchical Interface Graphics Standard
PID	Photo-imageable dielectric
PLCC	Plastic leaded chip carrier
PLD	Programmable logic device
PPM	Parts per million
PPO	Polyphenylene oxide
PPS	Polyphenylene sulfide (plastic)

Term	Definition
PRT	Planar resistor technology
PSI	Pounds per square inch
PT	Positional tolerance
PTFE	Polytetrafluoroethylene (Teflon)
PTH	Plated thru hole
PVC	Polyvinyl chloride
PWA	Printed wiring assembly
PWB	Printed wiring board
QFP	Quad FlatPack
QML	Qualified manufacturers list
QPL	Qualified products list
QTA	Quick turn around
R	Rosin (flux)
RA	Rosin activated (flux)
RAM	Random access memory
RETMA	Radio Electronics and Television Manufacturers Association
RFI	Radio-frequency interference
RFP	Request for proposal
RFQ	Request for quote
RFS	Regardless of feature size
RISC	Reduced instruction set computing
RMA	Rosin mildy activated (flux)
RMS	Root mean square
ROM	Read only memory
ROM	Rough order of magnitude
RPM	Revolutions per minute
RSS	Ramp soak spike
RwoH	Reliability without hermeticity
RTS	Ramp to spike
RTV	Room temperature vulcanizing
SAE	Society of Automotive Engineers
SBU	Sequential build-up

Term	Definition
SEC	Solvent extract conductivity
SEM	Scanning electron microscope
SEM	Standard electronic module (navy)
SERA	Sequential electrochemical reduction analysis
SFM	Surface feet per minute
SIP	Single inline package
SIR	Surface insulation resistance
SMD	Surface mount device
SMEMA	Surface Mount Equipment Manufacturers Association
SMOBC	Solder mask over bare copper
SMT	Surface mounting technology
SMTA	Surface Mount Technology Association
SNA	Systems network architecture
SOIC	Small-outline integrated circuit
SOS	Silicon-on-sapphire
SPC	Statistical process control
SPICE	Simulation program, integrated circuit emphasis
SQC	Statistical quality control
SQL	Structured query language
SSI	Small-scale integration (integrated circuit)
STEP	Standard for Exchange of Product Model Data
TAB	Tape-automated bonding
T's & C's	Terms and conditions
TCE	Thermal coefficient of expansion
TCP/IP	Transmission Control Protocol/Internet Protocol
TCR	Temperature coefficient of resistance
TDR	Time-domain reflectometry
TEM	Transverse electromagnetic mode
TFA	Tree-based floorplanning algorithm
TFE	Tetrafluoroethylene (Teflon)
Tg	Glass transition temperature
TGA	Thermo gravimetric analysis

Term	Definition
THT	Through-Hole technology
TIFF	Tagged image file format
TMA	Thermo mechanical analysis
TO	Transistor outline
TOP	Technical and office protocol
TP	True position
TQM	Total quality management
TTL	Transistor-transistor logic
UBM	Under bump metallization
UHF	Ultra-high frequency
UL	Underwriters laboratories
ULSI	Ultra-large scale integration (integrated circuit)
URL	Uniform resource locator
VAR	Value-added reseller
VHDL	VHSIC hardware description language
VHF	Very high frequency
VHSIC	Very high speed integrated circuits
VLSI	Very large scale integration (integrated circuit)
VME	Versa-module electronic
VOC	Volatile organic compound
VSAG	VHDL Standardization and Analysis Group (IEEE)
WIP	Work in process
WSI	Wafer-scale integration
WWW	World wide web
WYSIWYG	What you see is what you get
XIP	Execute in place
ZAF	Z-axis adhesive film
ZIF	Zero-insertion force
ZIP	Zigzag inline package

Index

About the Author

CHRISTOPHER T. ROBERTSON has worked nearly a decade in the PCB industry. His background includes developing standards and procedures for PCB design, creating a training course for new designers, and writing a regular column for a PCB trade magazine. Robertson has also worked in PCB manufacturing, testing, maintenance, assembly, and software beta testing and is also a member of the IPC Council.

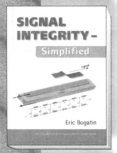

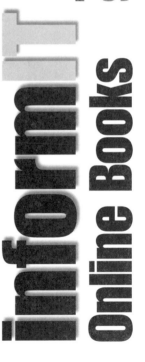

inform IT

YOUR GUIDE TO IT REFERENCE

Articles

Keep your edge with thousands of free articles, in-depth features, interviews, and IT reference recommendations – all written by experts you know and trust.

Online Books

Answers in an instant from **InformIT Online Book's** 600+ fully searchable on line books. Sign up now and get your first 14 days **free**.

POWERED BY
Safari

Catalog

Review online sample chapters, author biographies and customer rankings and choose exactly the right book from a selection of over 5,000 titles.

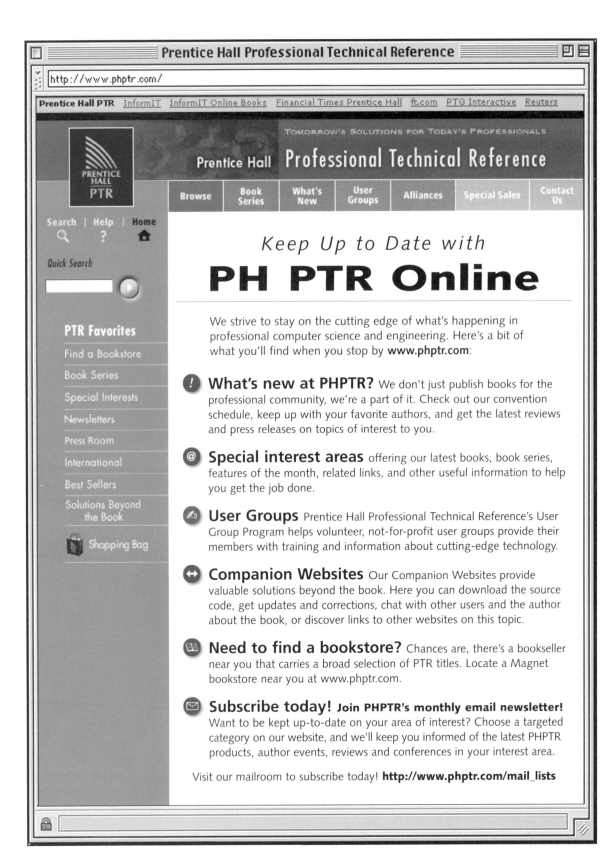

Company's only obligation under these limited warranties is, at the Company's option, return of the warranted item for a refund of any amounts paid by you or replacement of the item. Any replacement of SOFTWARE or media under the warranties shall not extend the original warranty period. The limited warranty set forth above shall not apply to any SOFTWARE which the Company determines in good faith has been subject to misuse, neglect, improper installation, repair, alteration, or damage by you. EXCEPT FOR THE EXPRESSED WARRANTIES SET FORTH ABOVE, THE COMPANY DISCLAIMS ALL WARRANTIES, EXPRESS OR IMPLIED, INCLUDING WITHOUT LIMITATION, THE IMPLIED WARRANTIES OF MERCHANTABILITY AND FITNESS FOR A PARTICULAR PURPOSE. EXCEPT FOR THE EXPRESS WARRANTY SET FORTH ABOVE, THE COMPANY DOES NOT WARRANT, GUARANTEE, OR MAKE ANY REPRESENTATION REGARDING THE USE OR THE RESULTS OF THE USE OF THE SOFTWARE IN TERMS OF ITS CORRECTNESS, ACCURACY, RELIABILITY, CURRENTNESS, OR OTHERWISE.

IN NO EVENT, SHALL THE COMPANY OR ITS EMPLOYEES, AGENTS, SUPPLIERS, OR CONTRACTORS BE LIABLE FOR ANY INCIDENTAL, INDIRECT, SPECIAL, OR CONSEQUENTIAL DAMAGES ARISING OUT OF OR IN CONNECTION WITH THE LICENSE GRANTED UNDER THIS AGREEMENT, OR FOR LOSS OF USE, LOSS OF DATA, LOSS OF INCOME OR PROFIT, OR OTHER LOSSES, SUSTAINED AS A RESULT OF INJURY TO ANY PERSON, OR LOSS OF OR DAMAGE TO PROPERTY, OR CLAIMS OF THIRD PARTIES, EVEN IF THE COMPANY OR AN AUTHORIZED REPRESENTATIVE OF THE COMPANY HAS BEEN ADVISED OF THE POSSIBILITY OF SUCH DAMAGES. IN NO EVENT SHALL LIABILITY OF THE COMPANY FOR DAMAGES WITH RESPECT TO THE SOFTWARE EXCEED THE AMOUNTS ACTUALLY PAID BY YOU, IF ANY, FOR THE SOFTWARE.

SOME JURISDICTIONS DO NOT ALLOW THE LIMITATION OF IMPLIED WARRANTIES OR LIABILITY FOR INCIDENTAL, INDIRECT, SPECIAL, OR CONSEQUENTIAL DAMAGES, SO THE ABOVE LIMITATIONS MAY NOT ALWAYS APPLY. THE WARRANTIES IN THIS AGREEMENT GIVE YOU SPECIFIC LEGAL RIGHTS AND YOU MAY ALSO HAVE OTHER RIGHTS WHICH VARY IN ACCORDANCE WITH LOCAL LAW.

ACKNOWLEDGMENT

YOU ACKNOWLEDGE THAT YOU HAVE READ THIS AGREEMENT, UNDERSTAND IT, AND AGREE TO BE BOUND BY ITS TERMS AND CONDITIONS. YOU ALSO AGREE THAT THIS AGREEMENT IS THE COMPLETE AND EXCLUSIVE STATEMENT OF THE AGREEMENT BETWEEN YOU AND THE COMPANY AND SUPERSEDES ALL PROPOSALS OR PRIOR AGREEMENTS, ORAL, OR WRITTEN, AND ANY OTHER COMMUNICATIONS BETWEEN YOU AND THE COMPANY OR ANY REPRESENTATIVE OF THE COMPANY RELATING TO THE SUBJECT MATTER OF THIS AGREEMENT.

Should you have any questions concerning this Agreement or if you wish to contact the Company for any reason, please contact in writing at the address below.

Robin Short
Prentice Hall PTR
One Lake Street
Upper Saddle River, New Jersey 07458

About the CD-ROM

The CD-ROM included with *Printed Circuit Board Designer's Reference* contains the following:

1. Designer's Checklist—MS Word and PDF versions of the Designer's Checklist contained in the book. The Word file may be edited for your use.
2. Resource Calculator—A bundle of calculators and tables for everyday use. Some of the calculators and tables are:
   - Stack-up calculator
   - Standard stack-ups
   - Copper thickness
   - Dielectric materials
   - Pad calculator
   - Class-to-class
   - Microstrip impedance calculator
   - Stripline impedance calculator
   - Asymmetric stripline impedance calculator
   - Dual stripline impedance calculator
   - Differential stripline impedance calculator
   - Differential microstrip impedance calculator
   - Embedded microstrip impedance calculator
   - Hole-to-trace
   - PCB standards
   - Clearance calculator
   - Screw hardware
   - Technology table
   - Electrical calculations
   - Amps-to-width & volts-to-clearance, including:
     - Trace width or current carrying capacity per trace width
     - Current division between traces
     - Trace clearance or voltage capacity dependant on internal/external use
     - Trace resistance dependant on trace width

- Available traces based on trace width and space available (for tight fits)
- Solder mask clearance calculated from annular rings and electrical clearances
- A quick table of current capacity per trace width and copper thickness is also provided.
- Wire gauge amperage rating
- Pattern naming scheme
- Package naming scheme
- EIA pattern table
- SMT land pattern creator
- Package codes decipher
- Mil resistor reference
- Lead bend table
- Display matrix
- Time
- Manufacturer contacts
- Links

3. Standalone Resource Calculator—A standalone calculator for material stack-up. Stack-ups may be saved and loaded into *.stk files, allowing the user to save desired stack-ups.
4. Start Files—A collection of boarders, details, charts, note and title blocks for the new designer. They are all in AutoCad format (*.dwg).

The CD-ROM can be used on Microsoft Windows® 98/NT®/2000/Me

LICENSE AGREEMENT

Use of the software accompanying *Printed Circuit Board Designer's Reference* is subject to the terms of the License Agreement and Limited Warranty, found on the previous two pages.

TECHNICAL SUPPORT

Any support information or questions should be directed to support@pcbdr.com.

Prentice Hall does not offer technical support for any of the programs on the CD-ROM. However, if the CD-ROM is damaged, you may obtain a replacement copy by sending an email that describes the problem to disc_exchange@prenhall.com.